IRON INDUSTRY AND METALLURGY: A STUDY

B. Sasisekaran

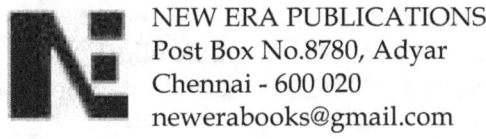

NEW ERA PUBLICATIONS
Post Box No.8780, Adyar
Chennai - 600 020
newerabooks@gmail.com

© *B.Sasisekaran*

First Published 2004
KDP Paperback Edition 2020

Published by : Srinivasan Srinivasan for New Era Publications,
Post Box No.8780, Adyar, Chennai-600 020. E-Mail: newerabooks@gmail.com

Dedicated to my father
late **SRI P.K. BALASUBRAMANIAN**
who was the essence of all that is good
and spiritual and is a guiding light to me

Contents

Foreword — v

Preface — vi

List of Line Drawings and Graphs — vii

List of Tables — vii

1: Introduction — 1-15
    1:1: Previous Research on Early Indian Iron — 01
    1:2: Distribution of Iron Ore in Tamilnadu — 07
    1:3: Iron Age Culture in Tamilnadu — 11
    1:4: Literary References to Iron Industry — 13

2: Pre - Industrial Iron and Steel Industry — 16-30
    2:1: Iron and Steel Production Centres in Ancient Period — 16
    2:2: Furnace Technology: Techniques, Types and Methods — 21
    2:3: Steel - Developments in Production Techniques — 25
    2:4: Production of wootz Steel by Crucible Process — 29

3: Metallurgical Studies of Iron Objects of the Iron Age — 31-59
    3:1: Typology of the Objects from Iron Age Sites in Tamilnadu — 31
    3:2: Guttur — 36
    3:3: Mallappadi — 38
    3:4: Kodumanal — 44
    3:5: Kanchipuram — 54
    3:6: Mel-Siruvalur — 57
    3:7: Explored Sites — 57

4: Summary — 60-63

Bibliography — 64-68

Select Index — 69

List of Plates — 70-72

SECRETARY
GOVT. OF INDIA
DEPT. OF OCEAN DEVELOPMENT
'Mahasagar Bhavan' Block-12, C.G.O, Complex,
Lodhi Road, New Delhi-110 003

DR. HARSH K. GUPTA   31 August, 2004

Foreword

The Monograph "Iron Industry and Metallurgy : A Study" in South India during Ancient times is a commendable effort by Dr. B. Sasisekaran. The study highlights the tremendous heritage of metallurgical skills of the Ancient Indian Communities since --700 BC. The author has undertaken a comprehensive study of the artefacts and furnaces found in the vicinity of excavations to unravel the nature and extent of iron smelting in ancient South India. The studies reveal the maturity of the ancient Indian metallurgist to obtain various categories of steel for different purposes. On the basis of the level of mineral transformation of iron artefacts, a hypothesis to relatively date the excavations has been expounded.

It may be inferred that the subsequent phases of alloy making using other metals had strong roots and foundations on Ancient Indian iron and steel making skills.

The monograph is an outcome of a painstaking review and focussed study of ancient habitations (excavations), for which Dr. Sasisekaran deserves appreciation.

H.K.Gupta

Preface

As far as I am aware, no one has attempted to embark on metallurgical study of iron objects of Iron Age, found in Tamilnadu and bring out a comprehensive monograph about the results of their research. Perhaps lack of motivation to visit normally inaccessible Iron Age sites and collect the iron objects and the absence of facilities to study them with sophisticated instruments with the guidance of experts may be the reasons for this. I believe that this lacuna is filled up to some extent with this publication.

This monograph is based on the original thesis of the post-doctoral research work done by myself when a fellowship was awarded (1994 to 1999) by the Indian National Science Academy (INSA) in New Delhi. I am grateful to the administrators and staff of INSA for their offer and help.

I remember here with gratitude the following scholars who guided me when the thesis was under preparation and thank them profusely for their help:

Dr. K.V. Raman, former professor of the Department of Ancient History and Archaeology, University of Madras co-ordinated my research work and provided the samples of iron objects and photographs;

Prof. R. Vasudevan, (Retd.), Department of Metallurgical Engineering and Material Science, Indian Institute of Technology (IIT), Chennai, initiated and guided me in to the field of Archaeo – Metallurgy;

Prof. B. Raghunatha Rao, Department of Metallurgical Engineering and Material Science, IIT, Chennai, became my co-ordinator in later period. The Metallurgical Studies on the iron artefacts included in the book were mainly due to the efforts of Profs.Vasudevan and Raghunatha Rao. I am very much indebted to the latter for the personal interest evinced by him in the progress of the work and his advice in the preparation of the draft of the thesis.

My sincere thanks are due to the authorities of the University of Madras, for permitting me to be a research associate in the Department of Ancient History and Archaeology and complete this post-doctoral research.

Scholars, technical personal and Institutions mentioned below helped me in various ways in shaping this thesis. I avail this opportunity to thank Prof. Kesavan Nair, Prof. K.J.L. Iyer (retd), Mr.T.Satyamurthi, Prof. Subbarayulu (retd), Dr. K. Rajan, Mr. Murugesan, Mr. Varadachari, Mr. Venkatavarathan, Mr. Seran, Dr. S. Rajavelu, Mr. G. Suresh Kumar, and the authorities of Madras Museum.

I am deeply indebted and grateful to Dr. Harsh K. Gupta, the Secretary, Department of Ocean Development, Govt. of India and an internationally renowned geophysicist, who in spite of his busy schedule has kindly consented to go over this monograph and contribute a learned foreword.

I may not have undertaken the publication of this thesis but for the encouragement from Dr. S. Kathiroli (Director), Dr. S. Badrinarayanan and my colleagues in the National Institute of Ocean Technology, Chennai. I thank them for their support when the manuscript was being revised for publication.

While converting a post-doctoral thesis into a book, meant for a wider audience, certain changes including the title of the book, became inevitable. Consequently minor additions and alterations were made, some chapters and paragraphs were shuffled and certain changes in the layout and numbering of the figures, tables and photographs were also carried out. But for these changes, the core of the thesis remains the same.

In this endeavour, I availed the expertise of Dr. Srinivasan Srinivasan of New Era Publications, Chennai. I am grateful to him for his devoted efforts and thank him for publishing this monograph.

B.Sasisekaran

List of Line Drawings and Graphs

1.	Map Showing Iron Age Sites in Tamilnadu	02
2.	Top View of Longitudinal Cross of Forge –Welded Iron Bar Section	39
3.	EDAX Analysis and Graph of Iron Bar	40
4.	Socketed and Barbed Arrowhead Showing Different Microstructural Features in Various Locations	45
5.	Leaf Shaped Arrowhead Showing Microstructural Details in Various Locations	46
6.	Line Drawing of Iron Chisel-Microstructural Details at Various Locations	48
7.	Line Drawing of Iron Nail - Microstructural Details at Various Locations	51
8.	X-Ray Difractogram of the Iron Slag – Kodumanal	55
9.	X-Ray Difractogram of Iron Tang – Perur	56
10.	X-Ray Difractogram of Fish Hook – Kanchipuram	58

List of Tables

Table – 1: List of Excavated Sites from where Samples were Analysed	35
Table – 2: Hardness Value of Various Iron Pieces in Various Regions	37
Table – 3: Micro Hardness of Iron Piece at Various Regions	38
Table – 4: EDAX Analysis and Graph of Iron Bar	40
Table – 5: Hardness value - Transverse Cross Section	42
Table – 6: Hardness Value - Longitudinal Cross Section	43
Table – 7: Socketed and Barbed Micro Hardness Values in Different Regions	45
Table – 8: Arrow Head - Hardness values at different regions	47
Table – 9: Iron Chisel - Hardness values at different regions	48
Table – 10: Dagger - Hardness values at different regions	50
Table – 11: Iron Nail - Hardness values at different regions	51
Table – 12: Iron Bead - Hardness values at different regions	53

1
Introduction

1:1: Previous Research on Early Indian Iron

The study of the Indian pre-industrial iron technology began around eighteenth century AD, when the westerners tried to understand the metallurgical properties and manufacturing process of the pre-industrial Indian steel known as wootz (Chakrabarti: 1992:1). The publication of Dr.Pearson's paper on wootz steel heralded the beginning of the metallurgical investigation on the pre-industrial Indian iron (Pearson: 1795:345). The term wootz for the Indian steel was originally derived from the Tamil word *urukku* (*Purananuru*: v.13) meaning fused metal or steel, also called *ukku* in Kannada and Telugu (Burrow: 1961:569).

The beginning of twentieth century heralded the study of iron in its antiquity, as against the earlier work on contemporary iron metallurgy. The scholars of this period researched on the origin of iron and its metallurgy from the literary and archaeological finds. While Schoff and Banerjee based their analysis on literary data, Bhandarkar and Marshall supplemented it by their report on the iron artefacts from Besnagar and Taxila, which contained the chemical and micro study analysis of Hadfield.

Dilip Chakrabarti divided the research on pre-Industrial Indian iron into four phases: 1) It centred mainly on the understanding of the process and properties of the Indian steel wootz. 2) This phase witnessed an attempt to study the composition of pre industrial Indian iron in relation to contemporary geological and metallurgical literature. 3) The third phase marked the beginning of interest in literary and /or archaeological data divested of the geological and metallurgical dimensions. 4) The above study led to the speculation on the origin and antiquity of iron in India with emphasis on literary data and correlating them with archaeological finds (Chakrabarti: 1983:81-82). The current research is on the distribution of pre-industrial iron producing centres, the availability of ore for pre-industrial smelting in different regions of India and outside on the basis of recent archaeological discoveries (Chakrabarti: 1992:1-22).

Begbie traces the antiquity of iron in India to c. 1100 BC, in his paper on the iron and steel industry in the Central Province. He opines that the Indians probably acquired the knowledge of iron from Assyria where it was produced around 1100 BC and the Indian metallurgists were well versed in the art of manufacturing iron on a large scale, centuries before the commencement of the Christian era. According to him the Indian wootz or steel was in great demand by cutlers (manufacturers of special surgical tools) in England.

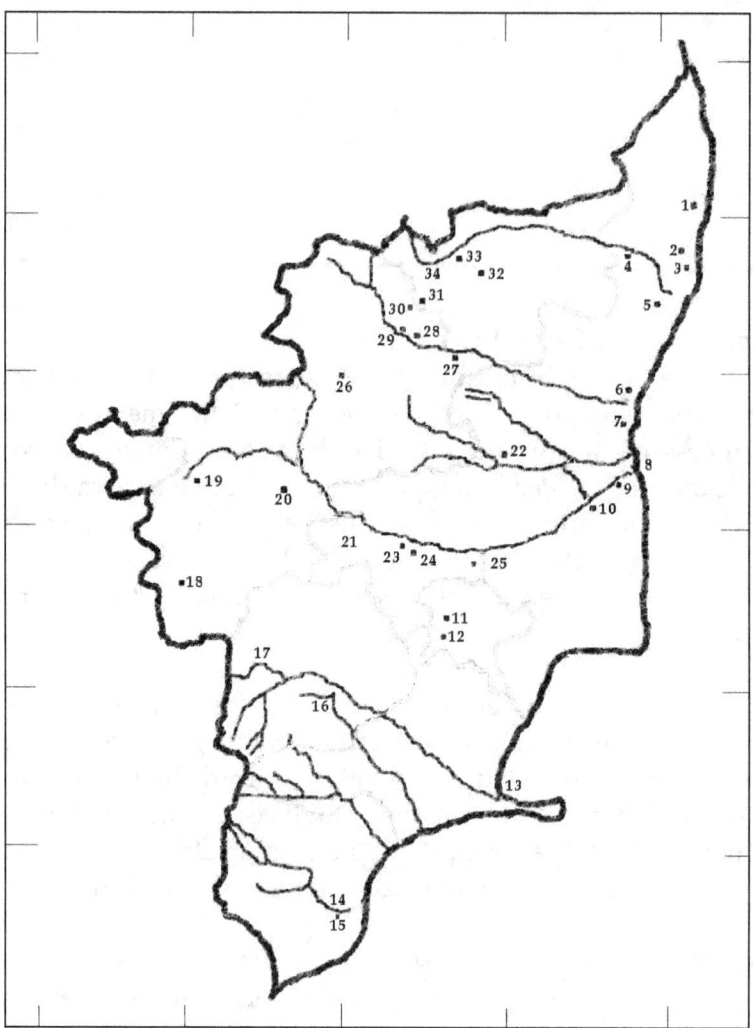

Fig 1: Iron Age Sites in Tamilnadu

1	Amirthamangalam
2	Sanur
3	Kunnattur
4	Kanchi
5	Perambir
6	Thiruvakkarai
7	Arikamedu
8	Kaveripattinam
9	Kambar medu – Terazendur
10	Vallam
11	Sittannavasal
12	Todiyur
13	Alagankulam
14	Korkai
15	Adichanallur
16	Kovalanpottal
17	Palani
18	Nattukkalpalayam
19	Perur
20	Kodumanal
21	Karur
22	Sengamedu
23	Tirukampuliyur
24	Alagarai
25	Uraiyur
26	Adiyamankottai
27	Mottur
28	Guttur
29	Togarapalli
30	Mallappadi
31	Paiyampalli
32	Odugattur
33	Appukallu
34	Kallerimalai

Its production was the cause of much wonderment and became the subject of various theories. The famous "Damascus Blades" which attained a reputation for flexibility, strength and beauty, were said to be made from the steel produced in an obscure Indian village. The Delhi Iron Pillar, the large iron bars found in the ancient temples of Orissa and the wootz steel show the metallurgical skill the Indian smelters achieved from early times, probably two thousand years ago (Begbie: 1908:1-2).

J.M.Heath, a Civil Servant in the services of the East India Company in the Year 1830, and its Resident Trade Representative at Salem, in his paper on the Indian iron and steel published in the *Madras Journal of Science and Literature* refers to the iron ore and the indigenous iron industry in the Salem district. Heath while tracing the antiquity of steel in India opines that the Hindus were the only people who were acquainted with the art of making steel. He further states that the Greek and Latin literature while discussing the quality and use of steel was silent on the method of manufacturing it from iron. Heath quoting Quintas Curtius, states that the gift of 14 kg of steel to Alexander by Porus was considered a present worthy of acceptance and according to him the steel must have been produced at a faraway place as the cost of the article had been enhanced by transport from the place of manufacture to the country of Porus (Heath: 1837:189).

Schoff in his paper on the eastern trade of the Roman Empire, with emphasis on literary data, states that the steel imported into Rome came from India. According to him some of the Indian steel might have been shipped through the Chera port but most of it went through the overland trade routes that traversed the dominion of the Andhra dynasty to the port of Barygoza on the Gulf of Cambay. These goods were carried westward by natives or Arabs and not by Romans (Schoff: 1915:236). The early Greek and Roman literature refer to the import of finest steel from India and the making of fine cutlery articles and armours at Damascus and Irenopolis (Banerjee: 1965: 161). Pliny refers to the swords made from finest Indian steel. Periplus of the Erythraean Sea, the work of an unknown author, mentions the import of Indian iron and steel into Abyssinian and African ports. The Periplus refer to the kingdom of the Chera as Cerobothra and their chief port, Muziris on the western coast. Periplus states that Muziri was an active centre of shipping from Arabia and Roman-Egypt.

Panchanan Neogi was the first researcher to make an integrated approach to the study of iron in India.In his work Neogi has analysed literary and archaeological evidence based on which he concludes that the early Aryan settlers in India were aware of iron. He did chemical and metallurgical analysis, studied the tensile strength, the forging and corrosion resistance of early beams and pillars from the archaeological finds such as the iron slag and clamps from Bodhgaya, spearheads and nail from Piprahwa, beams from Puri, Konark and Bhuvaneshwar, the pillars at Delhi and cast iron from Rangpur (Neogi: 1914: no. 12: 15-30). Though he refers to the iron objects from Adichanallur in detail, he has not given the chemical and metallurgical analysis of the iron objects from this place.

N.R.Banerjee discussed in detail the occurrence of iron in the early Indian archaeological context, in his monograph, *The Iron Age in India*, published in the year 1965.

Banerjee, based on his findings, concluded that the users of Painted Grey ware ceramic were responsible for the introduction of iron in the Doab region around 1000 BC and may have transmitted it by degrees to South India through the megalithic folks a little later (Banerjee: 1965:233).

Dilip Chakrabarti, in his account on the early use of iron in India, expounded the hypothesis of independent origin of iron in different regions of India. He observed that the recent archaeological data from South, East, North and Western India and Baluchistan on pre-industrial iron smelting, points to the earlier beginning of iron in Central and South Indian regions than in North and Western region (Chakrabarti: 1992:21). The evidence of pre-industrial smelting of iron is very impressive in Central and Southern India, which also seem to show the first evidence of Indian iron.

Pearson published his investigation on wootz steel in the *Philological Transactions* of the Royal Society of London. There he viewed that the wootz was made directly from the iron ore and was never converted into steel from wrought iron (Pearson: 1795:345). However, this view of Pearson was subsequently proved wrong by the memoirs of Francis Buchanan where he records in minute detail the process of smelting iron and making it into steel (Buchanan: 1991:285) and David Mushet, who pointed out that wootz steel was never made directly from the ore but by the fusion of wrought iron bars with woody carbonaceous matter covered with leaves and clay (Mushet: 1805:163-75).]

Buchanan, a surgeon by training, worked as an explorer and surveyor in the East India Company in India. He surveyed the pre-industrial iron industry of Mysore State and East India Company's acquisitions and provided a vivid account of the type of furnace used in smelting and the quantity of iron produced in each smelting operation in the regions of North Arcot, Coimbatore, Mysore, Canara and Malabar. Buchanan's memoirs were published in the year 1807 in three volumes. Buchanan found iron ore in all the areas he surveyed, forming beds, veins or detached masses in the stratum of indurate clay. It is full of cavities and pores and contains a large quantity of iron in the form of red and yellow ochre (Buchanan: 1807:Vol.II.436).

Dr.Voysey in his paper on the native manufacture of steel at Konasamudram situated about 20 km south of the Godavari and 40 km from Nirmul, mentions that the Deccani process involved the fusion of iron containing high and low carbon in the ratio of three parts to two inside the crucible. The crucible was made of granitic clay mixed with rice husk and oil. A small piece of glass or conch was put at the bottom of the crucible to serve as flux. The crucible was subjected to a constant heat for twenty-four hours and after that, it was allowed to cool. When cold, steel of great strength formed inside the crucible (Voysey: 1832:244).

Balfour, Surgeon General to the Government Museum, Madras, compiled, in the year 1855, an exhaustive list on the type of iron ore, the area where it was mined and the iron smelting centres that were still in active production in Madras Presidency. Balfour was the first scholar to make a systematic effort to document the distribution of iron ore, the cost of production and the value of iron in the market and prepare his report to the government (Balfour: 1855:4-5).

On iron industry in South India, Banerjee wrote that the rich iron ores of South India were responsible for the prolific and extensive occurrence of iron equipments in the megalithic and associated remains all over South India several centuries before the Christian era (Banerjee: 1965:190). Banerjee states that the furnaces in use in the South were circular on plan and conical in shape, broader at the base than at the top. They vary from 60 to 120 cm in height, about 25 to 40 cm across at the base and 15 to 25 cm at the top. There were two openings at the bottom, one for letting in the blast and the other for extracting the slag. Charcoal and ores were introduced from the top (Banerjee: 1965: 186). The wootz steel was produced in the Salem and Tiruchirappalli region by carburisation of wrought iron and the megalithic builders probably employed the same method of smelting iron ores (Banerjee: 1965:187).

Sarada Srinivasan from the Institute of Archaeology, London, in her paper on wootz steel in South India refers to the three regional variations in the production of wootz steel and named it as Deccani or Hyderabadi process, Mysore process, and Salem process. The Deccani process involves the fusion of high and low carbon iron with glass or conch as flux without any wood or carbonaceous matter inside the crucible. The Mysore and the Salem system differs in the method of cooling. In the Mysore process the ingot was fast cooled with water, but the Salem steel ingot was a slow cooled one taking as much as 24 hours to cool down to room temperature (Sarada Srinivasan: 1994: v.l.5: 51). She identified a wootz steel centre at Mel-siruvalur in South Arcot district with remnants of tapering tuyeres, furnaces, crucibles and blocks of slags from a mound covering an area of 25m x 9m x 5m. When reconstructed the fragments of crucibles showed typical features of the aubergine shaped closed crucibles used for wootz steel production (Sarada Srinivasan: 1994:54).

The early radio carbon dating of the level of Neolithic / Iron Age transition at Hallur, supported by the thermoluminescence dates from the neighbouring site Kumaranahalli (Karnataka) clearly strengthen the view of a pre –1000 BC beginning to iron in South India (Chakrabarti: 1992:80). According to him, the available records on pre-industrial smelting in Tamilnadu indicate considerable variation and interregional differences in the details of iron and steel manufacture. South Arcot and Salem districts show regional variation in the manufacture of wootz steel iron.

Buchanan refers to the smelting of iron at Venkatagiri, in North Arcot district, Virasolavaram on the southern bank of the river Noyyal near Chennimalai and Cootampalli near Coimbatore. According to him, "Shanar" community carried out the smelting operations in all these villages. They collected the iron bearing sand from the channels in rainy season, then they made charcoal, for using it in the smelting operation, from high-energy producing firewood like acacia and finally smelted iron in an interval of about three months between the crops season (Buchanan: 1988;Vol.I: 29).

Dr. Heyene surveyed the North Arcot district in the year 1818 for iron ore and the iron smelting industry in this area. In his report on the native iron smelting at Yrraguty near Sattur, Heyene states that the smelters collected the iron bearing sand from the small rivulets in the wet season or immediately afterwards and smelt iron in the dry season from the beginning of January to the end of March. They erected the furnace under the Banyan tree, which was exposed to the inclemency of the weather, without any other shade other than the shade of the tree. The furnace was made of red loam mixed with sand. It was made of two parts, the lower and the larger was about 22.5 cm in diameter, quite cylindrical and erected over a hole in the ground about 10 cm deep; its sides were uniformly 5 cm thick. The upper part was conical with the higher portion of the cone reversed. It was 45 cm in height and at the opening less than 30 cm in diameter. The bellows were made of sheepskin and a hole was left near the bottom of the cylindrical part of the furnace to fix the nozzle. The heat produced in this furnace was not sufficient to melt iron and the iron was spongy in condition and it required further heating and hammering to become perfectly malleable and fit for use (Balfour: 1855:14-15).

Salem and its surrounding neighbourhood had been the chief source of the supply of finest steel throughout historical period and probably from the earliest times. From the maritime intercourse between the Malabar coast, the Persian gulf, the stretch from the mouth of the Indus to the Red sea, it appears that the steel from South India found its way by these routes to the nations of Europe (Heath: 1837:190).

Captain Campbell, Assistant Surveyor General to the East India Company, refers to the presence of iron ore in large quantity in the granatic tract of Baramahal region. He constructed similar models of native furnace and tried to produce iron from them. On the basis of the results obtained by such experiments he concluded that a peculiar modification of the blast and proportion of the fuel, natural steel was produced as an immediate product. The economic cost of production and the welding properties of this natural steel made it immensely popular among the labour and poor class (Campbell: 1842:217-18).

The river Noyyal, a tributary of Kaveri, flows from the west to east, about half a kilometre south of Kodumanal. *Pathirruppattu* (v: 67,74) mentions that Kodumanal was an active trade centre trading in precious stones and other materials. The excavations at this place from 1986 to 1996 brought to light, factory sites where gemstone cutting, iron, and crucible steel (wootz) manufacturing industries were found in the bottommost level of the trenches. The archaeological materials revealed the beginning of industrial activity from the inception of cultural phase at Kodumanal (Rajan: 1994:1-2). In the year, 1987-88, American archaeologists from the University of Delaware led by Steven Sidebotham discovered two Black and Red ware potsherds with inscriptions in Tamil Brahmi characters at the site Quisel.E.L.Kadam near the Red Sea. The first sherd mentions the name of a person and the other to a merchant guild (Whitecomb: 1991:Vol.34.no.6).

The paucity of archaeological evidence to corroborate the early Greek literature led Schoff to think that the finest steel imported by the Romans came from central India. Two important archaeological discoveries, viz., 1) the discovery of two potsherds with Tamil Brahmi script near the Red Sea port, 2) the excavation of an industrial and trade centre at Kodumanal near Karur probably confirms the above mentioned hypothesis of Schoff. The Romans could have procured the steel from places like Kodumanal in Chera country in the South.

The smelters in the Salem region used the crucible process in which pieces of malleable iron were put in the crucible along with some wood and carbonaceous matter. The smelters in the Tiruvannamalai area manufactured steel by the decarburisation of grey cast iron. They made grey cast iron in a small blast furnace about three meters in height and used charcoal and ore, mixed in the ratio of 4: 1.The iron that separated was grey cast iron but without perfect separation. The grey cast iron thus produced was placed in a crucible with an equal portion of malleable iron produced from the bloomer (bowl) furnace. This type of manufacture was confined to a few families in the neighbourhood of Tiruvannamalai (Chakrabarti: 1992:145).

Rajan refers to the presence of iron slags, tuyeres, furnace material and Black and Red Ware and other associated potteries in the habitation mounds at Sulur, Nattukkalpalayam, Kaniyampundi, Kannarappalayam, Sirumugai, Idayapalayam and Kodumanal (Rajan: 1994: 93). The proximity of Nichampalayam and Idyapalayam to Kodumanal and the similarity of furnace material found in these sites to the one exposed at Kodumanal made Rajan to view that these sites were actively involved in the production of wootz steel in the early period. The occurrence of Roman wares, Brahmi letter bearing sherds reading Sanskritised names and punch marked coins show that the special steel from Kodumanal found its market in far away region (Rajan: 1994:89).

1:2: Distribution of Iron Ore in Tamilnadu

The systematic study on the distribution of iron ores in Tamilnadu had begun around AD 1800 when Francis Buchanan referred to the existence of iron ore in the Baramahal region. He was followed by a number of geologists who were collecting data on the presence of iron ores in the different regions of Tamilnadu. Notable among them were Dr.Heyene on North Arcot, Dr.Benza on the geological sketch of Nilgiri, Bruce Foote, and Alexander Primrose on the mineral potentialities in the erstwhile state of Pudukkottai, La Fennu, Thomas Holland, Dr. Dubey, Dr.M.S.Krishnan, and N.K.N.Aiyengar in the districts of Salem and Tiruchirappalli. The find spots of iron ore in Tamilnadu may be roughly divided into twelve regions. They are:

Salem district: The valley Thammampatti is a minor one.It is situated on the southern side of the Attur valley and is surrounded by the hills called Paiturmalai, Kollimalai, and Pachchaimalai. The rocks here are the gneisses with the general east-west strike and southerly dip.The indigenous smelting industry must have thrived in this valley. It is indicated by the occurrence of slags in heaps and pieces of crucibles scattered over a long distance near the village Sendarapatti. The rocks forming Taindamalai and Chitteri hills are similar to those in the Gudumalai and include gneisses, amphibolities, hornblende, chlorite, and talcose – schists and magnetite – quartzite. Thirthamalai and other associated hills are geographically the northern extension of the Chitteri and Kottapatti hills. There are four bands of magnetite – quartzites on this hill separated by gneisses and other rocks. Magnetite – quartzite are found in a number of places between Namakkal and Rasipuram. The length and thickness of the veins in these places vary from 2 to 15 km and 3 to 7 m respectively. Magnetite – quartzite is found in two bands in the Kollimalai area and in these, one is passing near Tottakkadu and the other to the east of Valavandinadu. Magnetite iron ore occurs in and around Kollimalai and the southern end of Pachchaimalai. A number of beds of iron ore in the Talamalai – Kollimalai group figure, very conspicuously in the great horseshoe curve of gneisses, in Namakkal, in the northwestern part of Tiruchirapalli district.

Nilgiri district: Iron ore occurs in abundance on the plateau and spurs of the Nilgiris. Magnetite is found in thick beds, either in the original rock or in the lithomargic earth. According to Benza the hematite ore present in the Nilgiris were originally magnetite altered by atmospheric agencies. Hematite is the next ore on the Nilgiris forming immense beds and sometimes, whole hillocks among the hornblende rocks and scientific granite (Benza: vol. 4.422). The hematite ore seems to contain feldspar, which on decomposing turns into yellow clay, lining some of the cavities in the rock. The next important group of iron ore occurs above the village of Kottagiri and a small spur of Dodabetta. At Jackatallapure, strings of hematite are found interfoiated with gneisses.

Chingleput district: The laterite of the Chingleput district is a ferruginous clay embedding small rolled pebbles and with all the characters of a sedimentary rock of the tertiary formation. The amount of iron in this clay varies in different localities. Brown hematite and magnetite is found mixed amongst the laterite between the Red hills and Nyour and they contain large and small pieces and blocks of quartz. Pieces of oxidulous iron ore are found over the surface of Pallavaram hill; on the sides and at the base of which black sand is exposed when it rains.

North Arcot district: Magnetite – quartzite ores similar to the one occurring at Kanjumalai of Salem district is noticed in a series of small hillocks near Manickavalli, Karapundi, and Nachamalai villages, in the Polur taluk. The State Geology Department estimated the iron ore around Polur to be of the order of four million tonnes with iron content of 38%. Several bands of magnetite – quartzite have been reported to occur in the Javadi hills. The lowest and the most prominent band is 30 m thick and has been traced from Merellimalai near Pudur upto Mathamalai, Southwest of Alangayam for a distance of 25 km. The ore deposit in the Kavuthimalai reserve forest area is in the two hills known as Vediappanmalai and Kavuthimalai located 9 km to the Northwest of Tiruvannamalai.

Coimbatore district: Coimbatore is a flat tableland averaging about 250 meters above the sea level. The district has on its borders the Nilgiris and Palghat ranges besides parts of Eastern Ghats like Anamalai, Sennimalai and other smaller hilly ranges inside its own territory. Ferruginous black sand bearing magnetite ore is found in the channels of little torrents, which wash it down from the hills in the rainy season.

Tirunelveli district: Captain Hoarsely and Bird report the presence of iron ore in the taluks of Nanguneri, Vayalur and in the Western Ghats. According to Bird the magnetite ore in the Western Ghat area is free of quartzite and contain mainly iron oxide and oxygen and the ore is darker in colour.

Madurai district: Iron ore is found distributed in the laterite conglomerate throughout the district of Madurai. Iron ore occurs near the base of the mountain and in places near Sivaganga and Cootampati.

Pudukkottai district: Magnetite is found in the gneissic rocks, about 2 km north-east of Malampatti. The laterite conglomerate in the Sengirai reserve forest area is thick and massive over an area of several square miles and is very rich in iron. The statistical account of the state (Pudukkottai) for the year 1813 mentions the presence of laterite conglomerate in large quantities at Andakulam, Perungalur, Thekkadu, Thiruvarankulam, Kilanilai and Sengirai. An inscription of Jatavarman Virapandya III at Tiruvarankulam dated 1299 AD referred to the smitheries and the profession tax levied on them (*Manual of Pudukkottai*, Vol. 1: 1938).

Tanjavur district: Iron ore is found in the laterite rocks of the western upland regions in the neighbourhood of Vallam and Gandharvakkottai. It is flaky ferruginous sandy clay, rather friable, but when exposed to the action of the sun and rain, it hardens and gets encrusted with dark polish of hydrated oxide of iron, which protects it from further change, and resists decay (Baliga: 1959).

Ramanathapuram district: Small pockets of magnetite occur in genesis, 2.4 km west of Elayiranpannai. The gneisses here show fine interbanding with magnetite.

South Arcot district: Iron ore is abundant in Kallakkuruchi and Tiruvannamalai regions of the South Arcot district. It also exists on the Kalrayan hills and particularly on the slopes below Chinnatiruppati, Ponparappai and Ravatnallur.

Let us now turn our attention to the type and quality of the iron ore found in different regions of Tamilnadu.

The richness of the Salem iron ore was discussed in an article. We learn from the *Journal of Royal Asiatic Society of Bengal* (Vol. I, 1832) that the iron ore occurs in the Salem region in the form of magnetic oxide and the hills in which it forms was found in the form of small grains, inter stratified with quartz and occasionally in regular octahedrons. The specific gravity of the octahedral crystals was found to be on an average 5.136 at 60°, which is rather more than is allowed in mineralogical works (1832: Vol.1. 253-55).

The magnetite-quartzite ore of the Salem –Tiruchirappalli – Arcot region constitute the most valuable group of iron ore deposit in Tamilnadu. The iron ore encountered in these areas is generally in the hilly tracts and they are divided into two equal parts by the Attur valley, which extends in an easterly direction from Salem towards Cuddalore (1944:

G.O. No. 4832). The northern area comprises the Shevaroy, Chitteri and Kalrayan hills, which spreads over a distance of 80 km from east to west. The northern extension of these is the Tirthamalai in which some important magnetite iron ore of good quality occur.

The best-known iron deposit is found in the Kanjumalai and Gudumalai region and both are comparatively small and isolated hills. The bands of ferruginous – quartzite's found in the Salem – Tiruchirapalli region are associated with a series of schistose and gneissic rocks, the most prominent among which are chlorite and tale schists and amphibole and garnet bearing gneisses. They form a conformable series, though metamorphosed and often folded. The metamorphism has been responsible for the conversion of the ferruginous sediments deposited, perhaps as mixtures of ferric hydroxide and silica into magnetite – quartzite.

The ores in these regions can be classified into three types. They are the banded, gneissic and fine grained. The banded variety consists of alternate bands of magnetite and quartz. The thickness of the bands and the proportion of magnetite to quartz in the different layers vary from place to place. The gneissic variety is akin to the banded one but the magnetite tends to be segregated into coarse lenticular or augen shaped grains. This makes the banding irregular. The fine-grained is composed of uniform mixture of grains of magnetite and quartz. When the grain is very fine, the rock appears aphanite and it requires the aid of a magnifying lens to distinguish the individual grains. A varying amount of hematite is present in most of the bands, quantitatively. However, hematite is unimportant as the average ore contains less than ten percent of the amount of magnetite present.

The iron ore deposits of Salem – Tiruchirapalli region are classified into the Kanjumalai, Attur valley Gudumalai area, Attur area, Tainandamalai and Chitteri hills, Tirthamalai, the area between Namakkal and Rasipuram, Kollimalai, Pachamalai and its spurs. The richness and availability of iron ore in the Salem region attracted the attention of scholars; notable among them are Sir Thomas Holland, Middlemiss and Dr.Dubey. Thomas Holland viewed the occurrence of hematite with magnetite in the Salem region a stage in the thermal metamorphism and the Kanjumalai area is by far the most abundant of the iron ores of the area. It occurs either in well defined octahedral crystals imbedded in chlorite schist or with quartz, a schist in which crystals of magnetite ore are crushed out in the direction of foliation to a roughly almond shape. Crystals of about one to three centimetres in length are common in the iron beds of Kanjumalai and other places in the Salem district. Magnetite also occurs together with small crystalline fragments of quartz, feldspar, hornblende, garnets and other minerals as sand in the riverbeds, derived from the disintegration of the numerous crystalline rocks within the area (Holland: 1892: *Records of GSI*, vol.xvii, 136-41).

Middlemiss states that the Kanjumalai is composed of great series regularly bedded gneisses, among which these containing hornblende, hypersthene, and garnets are conspicuous. The magnetic iron ore bands are not pure magnetite. The rock is a gneisses or schist, containing a very large percentage of quartz. The quartz – magnetite rock appear to

Introduction

be regularly found with the other gneisses. There are three such bands, the top band is a somewhat insignificant one forming a small eclipse of outcrop round the highest part of the mountain; the middle band about 150 m above the mountain from its base and about 10 m thick, composed of coarser quartz – magnetite rock. The lower most band about 20 m thick forms a well marked wall or dyke chain of spurs, almost all-round the mountain and which is slightly richer than the middle band (Middlemiss: 1896: *Records of GSI*, vol. 37-38).

Dubey viewed that the magnetite iron ore of the Kanjumalai contain iron and silica in almost equal proportion. According to him, they are in a series of five parallel bands totalling a thickness of about 100 m, running practically round the hill. He estimated the average amount of iron ore in the hill around 48% (Dubey: 1939:G.O.no.2872). Attur valley lies between Salem and Talaivasal as a long narrow strip extending to a distance of 65 km and 14 km wide. The gneisses including garnetferous, quartz schistose, chlorite schist's, amphibolitics and pyroxenites are the type of rocks present in the valley. The iron bed in the Attur valley takes its name from Gudumalai, a fine lofty hill rising in the middle of the great valley stretching eastward from Salem along the southern flanks. The Gudumalai divides the Salem –Attur valley into two parts, for it forms the watershed between the Vellar River and Tiruvanimuthur, a tributary of Kaveri. The magnetite – quartzite band of ore, which forms the central ridge of Gudumalai, runs east to west for nearly 5 km. According to La Fannu, a considerable number of native furnaces in the surrounding region of Gudumalai, received the iron ore for smelting operation in 1861 from the main bed at the west end of the Gudumalai ridge (La Fennu: 1883:98).

There are several hills other than the Gudumalai and Attur hills in the Attur valley between the Kalrayan and Kollimalai ranging in altitude between 300 to 600 m. The hills in this region formed mostly felspathic gneisses and amphibole rocks, with occasional roofs of magnetite-quartzite.

There are many references in the ancient Tamil literature, which indicates that iron industry flourished in Tamilnadu during the early centuries of the Christian era and they are provided below.

1:3: Iron Age Culture in Tamilnadu
The origin, antiquity, and the dispersal of iron technology in the Indian Subcontinent attracted the attention of scholars in the recent past. Mortimer Wheeler and D.H.Gordon held the view that the Indians learned the use and preparation of Iron from the Achaemenaids around 450 BC (Kosambi: *JESHO:* 1963:310-11). However, the excavations at Hastinapura, Alamgirpur, Ahichchhatra, Noh, and other Painted Grey Ware (PGW) sites pushed the antiquity of iron in north India to c.9th century BC. Chakrabarti identified six early iron using regions in the sub continent: Baluchistan, Northwest, Indo-Gangetic Divide, Upper Gangetic valley, eastern India, Malwa and Berar in central India and in the regions where megalithic monuments are found in South India (Chakrabarti: 1984: 82). Gaur expressed the view of diffusion of iron technology from West Asia. He states that the PGW using people probably acquired their knowledge of iron

from West Asia, especially from Iran, on geographical consideration and the use of iron from the Asia Minor region in the first half of the second millennium BC (Gaur: 1985: 80).

Iron, according to Chakrabarti, entered the Indian productive system by 800 BC and the central and southern India, with its rich iron ore and pre industrial smelting tradition, seem to show the first evidence of Indian iron (Chakrabarti: 1984: 83). But the evidence on iron obtained from megalithic sites in Deccan, Karnataka, Andhra Pradesh and Tamilnadu viz.; Naikund (BS: 265: 520± 100 BC, Habitation mound II, layer 6, Burial no.7, BS93: 545 ±105 BC), Takalghat (middle phase of megalithic habitation TF783: 615 ±105 BC & 555±100 BC) Bhagimohari (habitation –cum –burial, layer (9) BS 537: 690± 100 BC BS 536: 750 ± 100 BC) (Deo: 1994: 192), Hallur and Kumaranahalli in Karnataka and Veerapuram in Andhra Pradesh and Paiyampalli in Tamilnadu, indicate the diffusion of iron in Deccan, Karnataka and Andhrapradesh around 7th century BC and Paiyampalli in Tamilnadu around 6th century BC.

The Iron Age people, made their appearance during the late phase of the Neolithic culture i.e., c.600-500 BC, in the districts of Dharmapuri and North Arcot bordering Karnataka and Andhra Pradesh. The Iron Age of South India represents a distinctive phase of culture that succeeded the primitive neolithic culture. The new settlers by their knowledge of mining and metallurgy and their exploitation of rich natural resources enriched the pattern of living in the area of their settlement (Banerjee: 1965:208). The overlap of Neolithic – Iron age periods witnessed at Paiyampalli in Tamilnadu has also been observed at Hallur (*IAR*: 1964-65:31-32) in district Dharwar, Banahalli (*IAR:* 1983-84:42-46,*IAR:* 1985-86) in district Kolar in Karnataka and Hullikalu and Pagidigutta in Andhrapradesh (Chakrabarti: 1984: 84) The excavations at Banahalli have provided a clear cut picture about the developmental stages of the transition from Neolithic to iron age (Dikshit: *Puratattva*: 22). The Iron Age in Tamilnadu had a short span of time and ended with the beginning of early historical period in the late centuries BC and the early centuries AD. But the practice of erecting memorials i.e. Hero- stones continued longer and even up to the mediaeval period (Chakrabarti: 1992:80).

The settlement pattern of Iron Age habitation sites showed the preference of the authors of Iron Age for perennial rivers or their tributaries and in the absence of major river system, they made their settlement near perennial ponds. The Iron Age habitation in Dharmapuri and North Arcot region reveals their concentration along the course of river Pennaiyar and its tributaries. The habitation sites situated near the tributaries of Pennaiyar include Guttur, Mallappadi, Togarapalli, Dailmalai, Mullikkadu, and Chandrapuram in Dharmapuri district, Paiyampalli, Kallerimalai, and Chengam in North Arcot district. There is a wild stream running near the habitation site at Appukkallu. The Iron Age folks at the time of their entry into these regions appear to have followed pastoral economy and moved from place to place. The migration from the point of entry to the interior region probably showed their progress they achieved from village-based economy to the establishment of large towns, where trade, commerce, and industry could have flourished.

The presence of metal artefacts, especially of iron, besides furnace materials and enormous quantity of iron slag in the lowest stratum of the Iron Age settlements indicate the presence of artisans in substantial number in the community (Deo: 1983: 91). The large number iron tools of offence recovered from the burial and habitation sites must have kept the artisan group busy the whole year for the supply of these artefacts (Deo: 1983: 92). The typology of megalithic iron artefact indicates their use in stone – quarrying and wood working tools, arms and weapons and agricultural tools (Chakrabarti: 1992:81).

The discovery of iron and steel furnaces at Guttur, Kodumanal and other places reveal the existence of a class of artisan with professional skill and expertise within the megalithic community. The twin furnaces exposed at Guttur, iron slag and artefacts found at Paiyampalli and Appukkallu, datable to c.500 BC, and the iron and crucible furnace at Kodumanal, c.300BC point to the advanced state of technical knowledge of the people of this region in iron smelting. This also indicates that the early migrants of Iron Age were not incipient in their knowledge about iron smelting and they might have acquired it somewhere outside Tamilnadu, on their way before entering Tamilnadu. The Neolithic-Iron Age overlaps in phase I at Hallur (T.F573: 955±100 BC) and Kumaranahalli (PRL: TL: 50:1140 ±270) in Karnataka and Veerapuram (PRL 728: 920 ± 140 BC and PRL 730: 1200 ± 140 BC) in Andhrapradesh have been dated to c.1000 BC on the basis of a C-14 date (Deo: 1994:193). However, Iron made its appearance in the middle levels of period II at Hallur (Gaur: 1983:79-80). Based on its occurrence in the middle levels of period II at Hallur, the introduction of Iron in productive system in Karnataka region can be safely deduced to around 700 BC. The districts of Coimbatore, Dharmapuri, and North Arcot bordering Karnataka and Andhrapradesh, formed the nuclear zone of Iron Age culture in Tamilnadu. The iron age culture diffused towards the other parts of Tamilnadu from the point of entry in these regions.The iron age settlements are riverain in nature and they diffused towards the interior of Tamilnadu through the course of the rivers.

1:4: Literary References to Iron Industry

The early Sanskrit literature *Yuktikalpataru* and *Rasaratna - Samuchchaya* mentions about the m*unda* (cast-iron), *tikshna* (sharp iron or steel) and *kanta* (wrought iron)(Neogi: 1914:45). At present three principal varieties of iron are recognised according to the composition carbon in it viz. wrought iron steel and cast iron.

Early Tamil literature, popularly known as Sangam literature, mentions various kinds of iron and their different properties. From this literature we learn about iron viz.; *irumbu* (wrought iron) (*Purananuru*, v.170) *urukku* (steel) (*Purananuru*: v.130 *irumbu* (cast iron) (*Kurunthokai*: v.155). *Ahananuru* mentions iron and steel by the term *Irumbu* (v.72.2-6) and *urukku*. *Purananuru* differentiates ordinary iron from steel by emphasising the superiority of weapons made of steel (*urukku*) (v.13) and weapons made of steel (v.61-13, 304.4-5) were always referred to as "*ekku* or" *ekkam*". The data from Sangam literature was used by Vaithilingam (1977:268) in his monograph on fine arts and crafts in *Pattuppattu* and *Ettu-t-toki*. He vividly illustrates the black smith forge, iron industry, the furnace, the hand worked and pedal type bellows, the blowpipes, the anvil, sledgehammer and the tongs.

Apart from the references to iron, steel and black smith forge mentioned above, the Sangam literature abounds with reference to the iron smelting operation and the importance of ironsmith in the ancient war loving Tamil society. Different class of people were engaged in smelting of iron from ore and converting it into steel and manufacturing of weapons from it. The relative position of the ironsmith in the society is mentioned in *Purananuru* (v.287.1-3, v.170). People of low caste carried out the iron smelting operation and the black smith converted them into steel and produced other value added products.

The excavations at Kodumanal brought to light, the living quarter of the iron smelters, which was a simple structure with mud flooring and it was on the periphery of the habitation area, while the well-paved floor of the steel manufacturers was found in the midst of the habitation area. This information tallies with the description of the working place of these two types of artisans found in the Sangam literature.

The blacksmith formed an integral part of the early historic society, assignable to the first century BC to 3rd century AD. He was a state sponsored worker in the armies of the Tamil kingdoms. *Purananuru* (v.268) states that it is the duty of the blacksmith to manufacture *vel* or dart for the gallant soldiers.

The blacksmith, his forge and instruments are also referred to in the classical literature in different contexts: The *kollan* or iron-maker or iron-monger cum blacksmith; *Karumkaikollan*, the skilled worker in iron; *ulai* or his furnace; *turutti* or *visaiturutti*, the hand worked or pedal bellows; *kuruki*; the blow pipe or nozzle; *ulaikkal*, the stone anvil (Pl.1:1); *kudam*, the sledge hammer; and *kuradu*, the tongs are mentioned in the classical literature (Kuppuram: 1989:259). The blacksmith's forge with all its equipment is mentioned in *Ahananuru*. (v.202.3-8). It further refers to the sparks flying off the blacksmith furnace (*ibid, v.72.2-6*) and the fire in it blown through the blowpipe or nozzle (*ibid, v.224.2-5*). The *Perumpanarruppadai* (v.207-8) and *Narrinai* (v.125.1-5) mention the operation of blast furnaces by the pedal bellows. From *Purananuru* (v.170.15-16), we learn that the black smith operated the bellows with an assistant who repeatedly tread on it with his foot. We also understand that the anvil (Pl.1:1) (*ulaikkal*) receives all the shocks when the powerful hammer strikes the object of work on it

We are also told that while preparing the wrought iron from the bloom, sparks flew in all directions when the smelter hammered the red-hot iron on an anvil to remove the non-metallic inclusions (*ibid*, v.202.3-8). Iron smelted from the ore was a pasty semi-solid mass with lot of non-metallic inclusions. The iron thus produced required further treatment to remove non-metallic inclusions. Hammering the red-hot sponge iron on an anvil did this. When hammered, the sparks flew in all directions from the sponge iron, and when sparks ceased, the blacksmith knew that the metal had become homogeneous.

Purananuru also refers to the quenching of iron piece held by the tongs and heated red first in the fire and then plunging into the water (v.21.7-8). Other than the weapons of offence and related articles (the ferrules covering the tusks of elephants are generally made of iron) objects of daily use like knife and chisel were all made of steel. Another classical literature *Kurunthokai* refers to the artefact made of iron by casting. It mentions that iron lamps and bells were made by *cire - perdue* (lost wax) process in the blacksmith foundry (v.155). The excavations at Guttur brought to light, an iron foundry and the iron artefact made of cast iron from those place as early as c.500 BC. The recovery of hollow terracotta ring mould with a hole or spout like thing from iron smelting industrial sites at Nattukkalpalayam and Kannarappalayam for metal casting further support the early literary reference on iron artefact produced in ancient Tamilnadu by metal casting (Rajan: 1994:97). These terracotta rings might have been used to make cast iron rings. The molten iron was poured inside the mould through the hole or spout and the mould were splashed with water for fast cooling, as evidenced from the analysis of artefact from Guttur (Raghunatha Rao, *et.al.*, I.N.S.A: 32(4) 1997:354). From the cooled mould the metal was removed by breaking the terracotta ring.

The excavation (period II, 500-200 BC) at Ujjain brought to light, remains of a Blacksmith forge consisting of all the necessary equipment such as a groove for the introduction of nozzle of a bellows, an improvised stand made from the large of a broken vessel to support a water jar to store water for quenching, the use of an anvil and iron tools like pincers (Banerjee: 1965:179). Though Ujjain is far away from the Tamil land, early Tamil literature, *Manimekalai* states that the black smiths from Ujjain were known and sought after by the early Tamil people. (*Manimekalai*: XIX, II, 107-9).

2
Pre - Industrial Iron and Steel Technology and Industry

2:1: Iron and Steel Production Centres in Ancient Period

The vestiges of iron smelting operation dating back to 8th - 7th centuries BC was brought to light, in the form of furnaces and vitrified iron slag from archaeological excavations from ten places out of which seven are from Tamilnadu. Notable sites where iron and steel smelting furnaces were found include Jodhpura (Agrawal & Vijaykumar: 1976:242), Naikund (Gogte: 1982: 11), Dhatwa (Mehta, *et. al.*, 1975: 40), Atranjikhera (Gaur: 1985:79), Ujjain, (Banerjee: 1965: 178), Guttur (Raman: 1983:67), Kodumanal (Rajan: 1994: 93). A brief account of the furnaces found in some such places is given below.

Jodhpura: The discovery of two iron-smelting furnaces in early phase of PGW (c.750 BC) at Jodhpura (1976, District Jaipur) points to an early tradition of iron metallurgy in India. The furnaces were of open type and provided with bellows thereby indicating the advanced technique employed in smelting. A preliminary examination of the available evidence shows that the iron smelters at Jodhpura exploited iron ore and fuel resources available locally and probably extracted iron directly in a primitive furnace without fluxing the ore. Of the two furnaces exposed at Jodhpura, one was used to extract bloom and second hearth was meant for heating during forging. A platform between the two hearths was provided. The smelters probably used the platform to forge the bloomer after it was taken out from the furnace and heated red-hot in the open hearth (Agrawal & Vijaykumar: 1973:242).

Naikund: The excavations (1982) at the megalithic site Naikund (21°20'N 79°10'E) near Nagpur revealed an iron-smelting furnace made of clay, circular in plan with clay tuyeres and an opening for easing out tapped slag. The diameter and the height of the furnace were 30 cm and 25 cm respectively. Circular clay bricks were piled one over the other; the upper surface of the lower brick was convex and the lower surface of the upper brick was concave. The cross section of the circular bricks showed that the inner side, of about 2 cm thickness was turned black due to firing in reduced conditions, while the outer side was brownish red. A hole at the bottom was provided for the tapped slag. A few bricks, forming the bottom of the furnace, were found fused with slag and cinder (Gogte: 1984:545). Gogte states that the furnace at Naikund in a single operation could turn out nearly 3 kg of pure iron out of about 10 to 12 kg of ore (Gogte: 1984:89).

Dhatwa: Dhatwa (21°18' N, 73° 04' E) lies on the southern bank of the river Tapi in Surat district. The excavations in the industrial site at Dhatwa revealed evidence on iron smelting operation in the fourth and third centuries BC, in the form of iron objects, heaps of tapped slag, iron ore pieces in the stratified layers and on the surface of the mound. Hegde on the basis of his analysis of furnace and other associated materials concluded: the iron smelting and smithery started at Dhatwa as soon as the early historic community settled at the site; More number of iron objects and slag's and pieces of ore were recovered from layers 2 and 1 than from layer 3; The presence of slags on the surface of the mound indicates the continuing industrial activity as long as the early historic community lived at Dhatwa; The quantity of slags recovered from stratified layers show that the iron smelting industry was fairly large; The iron objects recovered in the excavation reveal their use for agriculture in the fertile Tapi valley. Analysis of slag samples from Dhatwa revealed much of the iron in the ore was lost in the slag, as the ore was smelted at a low temperature of reduction of about 1100°C to 1200°C. The extracted bloom must have been pure and comparable to the present day wrought iron. The chemical analysis of the metal sample from Dhatwa shows that the metal was remarkably pure (99.76%) and this purity was found in a forged object, which is likely to be carburised (Hegde: 1973:403). Metallographic studies of iron hoe from this site indicate that it was made in two stages. First the red-hot bloom was forged into thin sheets on an anvil near an open-hearth. In this process the surface of the sheets was carburised and casehardened. Secondly, several of these sheets were joined laterally, one by one, by forge welding. Finally, the whole mass was further forged to shape it into a hoe. The laminated structure in the metal indicates the lines of lateral welding (Hegde: 1973:404).

Guttur: The excavation (1982-1983) at Guttur (12° 25' N 78° 15' E), hereafter GTR, near Krishnagiri brought to light, for the first time in Tamilnadu an industrial centre where iron artefacts were produced from around c.500 BC. The surface around the foot of the hill contains large number of iron slags, cinder, blowpipes, tuyere pieces with vitrified mouth and ashy white soil, indicating flourishing industrial activity from early period c.500BC (Pl.1:2) The dig in the trench GTR IV, exposed a twin elongated oval furnace, each measuring 2.02 m in length, the width and the depth of the furnace measures 0.63 m and 0.45 m, the thickness of the wall portion measures 0.04 m on its northern side and 0.08 m on its southern side (Pl.1:3). Brick structure was found on either side of the furnace and in between the twin furnaces. The one at the middle was probably used for the bellows and the brick structure on its side for the filling of the furnace with fuel and ore while the smelting is in progress. The exposed portion of the furnace showed three openings, one on its side and two in front with earthen pipes (Pl.2:1). The one at the bottom indicates the arrangements made for the retrieval of molten iron on its sides (Pl.2:2). One of the other two openings near the top was for the bellows and for removing the slag from its front side. Fine Black and Red ware sherds, iron artefact and iron slags were recovered in stratified layers along with exposed portions of the furnace. The measurement of the furnace showed that it could have been the largest furnace in operation at that time in India. Prakash quoting Buchanan states that the twin hearth furnace of Malabar was the largest furnace with a production capacity of 250 kg per heat. According to him the furnace came into operation only in the 18th or 19th century AD (Prakash: Puratattva: 1989-90:No20: 119). However, the excavation at Guttur revealed the existence of similar twin hearth furnace in Tamilnadu as early as c.500 BC (*IAR:* 1982-83:67).

The smelters in the Malabar region also used a twin furnace in their smelting operation. The front of this type of furnace was quite open. The production process used in these furnaces yielded only 11% to 18% of iron. The main defect in the process seems to be want of proper bellows (Buchanan: 1988: Vol.II: 436,439).

The chemical and metallurgical analysis of the iron artefact from Guttur showed that it was a cast iron with carbon content varying from 3 to 5%. The question of whether production of cast iron at so early a period viz., c.500 BC, at Guttur in Tamilnadu was a deliberate one by the ancient metallurgist in India or accidental is answered by the discovery of hollow terracotta rings in the course of archaeological excavations at Kannarappalayam (Stanford: 1901:461-471) near Mettuppalayam and Nattukkalpalayam (Walhouse: 1875:17-34) datable to iron age period (c.5rd century BC). The terracotta pipes measured 30 cm in diameter and was provided with a spout. The molten iron was poured inside the ring through the spout and left for sometime. On cooling, the iron would get the shape of the terracotta ring and the cast iron recovered by breaking and then opening the mould. The terracotta rings recovered during the course of the archaeological excavation in these sites might have been used to make circular rings or strings (Rajan: 1994:97).

The terracotta ring mould unearthed in the excavations in Coimbatore district and the discovery of twin furnace and cast iron artefact at Guttur c. 5th century BC, indicate the existence of iron foundry, where the manufacturers produced finished artefacts such as iron rings, iron bells etc., by casting, as manufactured in an iron foundry in modern times. The manufacture of iron artefacts in ancient times by *cire-perdue* process has been referred to in the Sangam anthology, *Kurunthogai*, which states that iron bells were manufactured by casting, a process the early iron metallurgist inherited from bronze metal casting (*v*.155).

In addition, the discovery of a furnace in GTR IV, the excavation in trenches GTR I and GTR II revealed structures contemporaneous to the smelting operation at Guttur. The presence of postholes indicates that the structures were covered with thatched roof. They were probably used as shelter or storage yard for finished goods.

Kodumanal: The excavations at Kodumanal (1985-1996) (11° 6' 42" N 77° 30' 51" E) yielded material evidence on the industrial activity carried at that site from 300 BC (Pl.3:1) The excavation in one of the trenches revealed a dismantled bowl furnace at a depth of 65 cm right on the natural soil.The shape of the furnace was distinguished by burnt clay having smooth surface and circular form with white colour caused by high temperature. The base of the circular bowl furnace measures 115 cm in diameter. The height of the furnace could not be ascertained since it was recovered in "dismantled" (broken) form. However the measurement of the base of the furnace indicates that it was bigger than the bowl furnaces that were in use at Salem and Coimbatore regions in the 18th and 19th centuries AD. The height of the furnace should be correspondingly higher than the 19th century furnaces in this region and 4 feet tall (120 cm). Brick pieces in vitrified form indicate their use, though in limited number, in the construction of the furnace. Ashy patches and charcoal noticed near the base of the furnace indicate their use in smelting operation. Tuyeres measuring 15 cm in length, 6 cm in thickness and 1.5 cm in diameter with vitrified mouth indicate that the bellows were used quite close to the furnace. Some of the iron slags stuck to the wall portion of the furnace had a smooth surface. The presence of broken furnace materials in the other two trenches laid near the former, indicate flourishing industrial activity at Kodumanal. A wrought iron piece was found near one of the broken furnaces indicating the kind of iron produced at Kodumanal (Rajan: 1994:94). Bowl furnace used at Kodumanal could attain a temperature of around 1180°C and iron recovered at the end of the operation was spongy and semi-solid with lot of slag and other impurities embedded in it. It requires further heating and forging before it was made malleable and soft for further use. The granite slabs found near the furnace suggest their use as an anvil for forging. The absence of post-holes, floor levels, less occurrence of potsherds in the smelting area indicate that the smelting of iron from ore was carried probably to avoid pollution on the periphery of the habitation.

Mel-Siruvalur: Mel-Siruvalur is situated 8 km south of Moongilthuraipatu on the Sankarapuram road in Sankarapuram taluk in South Arcot district. The megalithic cairn circles are found on the western side of the road in disturbed condition indicating the occupation of the site in antiquity.

According to the Manual of the South Arcot district, 1878, iron is found in abundance in parts of Tiruvannamalai taluk and on the slopes of the Kalrayan hill below Chinnatiruppati, Ponparappai, and Ravatnallur in Sankarapuram taluk. Three new industrial centres datable to early historic period (c.200 BC to 300 AD) were found at Pakkam (Pl.3:3), Puthianpettai and Ravatnallur besides Mel-Siruvalur. The antiquity of the sites mentioned above is indicated by the surface collection of Black and Red ware, Black ware, and Red ware potsherds and the discovery of steel ingots (Pl.3:2), and vitrified wootz crucibles. Other than these finds two Hero stones datable to Pallava period were found at Pakkam and Mel-Siruvalur. Numerous crucible fragments and bowl shaped vitrified crucibles in broken condition together with fragments of glassy slag charge were collected in a field south of the habitation. The crucible bowl measures 18 cm in diameter (Pl.3:4). The presence of vitrified slag, broken crucibles and debris point to the functioning of steel industry in early times, since the villagers had no memory of any metallurgical activity in recent times in this village. The industrial site covers an area 25 m wide and 5 m height (Sarada Srinivasan: 1994:52).

The furnace area at Mel-Siruvalur was demarcated by a set of two trenches inter connected in a pitch and swell shape of about 10 m long. It contained several tapering tuyeres fragments (with an inner diameter of 1.5 cm, and varying from 0.8 - 2 cm thick); along with furnace remnants consolidated mud and slag (Pl.3:5). The other trench contained only blocks of slag 20 cm high and 20 cm in diameter with a flow structure; indicating that it had been used to tap out slag from the main furnace (Sarada Srinivasan: 1994:54).

The broken crucibles and some of slags found in the trench nearby were analysed by Sarada Srinivasan. The investigation indicated that the fabric of the Mel-Siruvalur specimens consisted of a porous glassy matrix with distinctive cooked rice hull relics dispersed in the matrix along with sand or quartz grains - a distinctive feature of the manufacture of Deccani wootz crucibles. The slag from the trench on analysis showed that the main constituents were iron and silica suggesting that these were of fayalite type. This suggests the smelting of iron in a bloomer furnace. The iron bloom produced may have formed part of the charge to produce wootz steel by crucible process (Sarada Srinivasan: 1994:56).

Kattankulathur: An early iron-smelting site is found at Kattankulathur, about 35 km south of Chennai and 5 km north of Chingleput town on the national high way. The site is located 1 km southeast of Katankulathur railway station. The industrial site measures an area of 10 acres with two mounds in the middle. The mounds measure 500 square meters; one is partially exposed revealing the base of a furnace, tuyere, blowpipes, and slag. Black and red pottery pieces were picked while clearing the vegetation around the furnace (Pl.3:6). Two round pits each measuring 45 cm in diameter were found scooped out of the natural bedrock near the mound (Pl.4:1). The smelters probably used the pits to pound the ore into small granules. The account of Balfour in the year 1855 refers to the presence of iron bearing rocks 5 km south of Chingleput town in the Venpakkam hills. The ore in the Venpakkam hills might be the source for the iron smelters of Kattankulathur. Wet chemical analysis of the iron slags recovered from the furnace area show high Manganese 0.58%. Phosphorus and sulphur is low at 0.38% and 0.19% respectively. High manganese slag obviously comes from high Manganese ore. Manganese oxide present in the ore acts as flux and helps in the reduction of ore into metallic iron.

Perungalur: Perungalur (10° 56′ N 78° 29′ E) lies 30 km from Thanjavur towards Pudukkottai. The statistical account of the then Pudukkottai state for the year 1813 refers to the rich iron bearing laterite deposits and iron smelting operation at Perungalur. The antiquity of the town is seen by the presence of megalithic burials datable to 1st century BC. In the course of the author's exploration, a bowl shaped iron furnace along with slags, blowpipes, tuyere and Black and Red ware, Black ware, and Red ware pottery pieces were found in a farmer's field (Pl.4: 3). On enquiry the furnace was found unearthed along with the other materials three feet below the ground in the midst of the field while the farmer was digging for a well. The site where the furnace was discovered lies on the outskirts of the Perungalur town towards Thanjavur. The furnace was elliptical in shape and measures 65 cm in width and 55 cm in breath. Wet chemical analysis of the sample from Perungalur showed that the product was low in carbon and probably an attempt to make steel by direct method.

Tiruvarankulam: A Pandya inscription from the temple at Tiruvarankulam town (10° 56′ N 78° 22′ E) dated in the 4th year of Virapandya assignable to 13th century AD, speaks of iron smelting industry in this town and the black smiths colony, on the western side of the temple. The inscription also mentions the tax levied on the furnace in operation at Tiruvarankulam (*Manual of Pudukkottai*: Vol. I: 1938:196). Though the present generation does not know much about the iron smelting carried on in earlier times, the author discovered two round furnaces about 1 km north-west of the Kammala colony. The furnaces were found embedded on the banks of a tank formed by the excavation of soil around the place for brick industry.

Veppangudi: An iron-smelting site is found near the village Veppangudi about 3 km east of Tiruvarankulam. Iron furnace blowpipes and tuyeres are found embedded with slag forming into a huge mound covering an area of one-acre (Pl.4:4). The wet chemical analysis of slag from Veppangudi showed high Manganese as at Kattankulathur in Chingleput district. A megalithic habitation site is found near the village Ponparakkottai. The habitation site is situated 4 km from Tiruvarankulam. The habitation mound rising to a height of 30 feet from the adjacent lake covers an area of 20 acres. At present the mound is under cultivation except for a small pocket of about fifty meters, left barren on either side and from where the author was able to collect iron slags, blowpipes, tuyeres and Black and Red ware, Red ware and Black ware ceramic from the surface. The industrial site near Veppangudi and its relation to the other two sites viz. Tiruvarankulam and Ponparakkottai requires further investigation.

2:2: Furnace Technology: Techniques, Types and Methods

In the history of metallurgy, Iron came late because of the difficulty encountered in the technique to smelt ferrous ore compared to the more common non-ferrous ores. Though the melting point of iron is 1540°C, it can be reduced to pure iron oxide at about 800°C, considerably below its melting point (Tylecote: 1964:183-84). Iron technology developed in India on the basis of long-standing copper smelting tradition. The knowledge acquired in smelting copper over a long period from c.3000 BC provided the necessary infrastructure for mining, roasting and crushing the ore and its smelting in a small simple furnace (Joshi: 1970: 38). The technique required for smelting iron is a little more difficult than the one required for copper. Affinity of oxygen to iron is much stronger than that to copper and the Iron ore consists of not only Iron oxide but also more gangue (valueless and undesirable materials in ore) than copper ore. Since it is not practicable to remove all the gangue by washing, slagging in the smelting process must separate a large amount of it. Therefore, it requires a condition that is more critical for smelting. Iron requires a minimum temperature of 1250°C, to remove gangue from the smelting charge. To attain this temperature a good supply of oxygen to the furnace is needed. With so much supply of oxygen it is difficult to maintain reducing condition in the furnace. To offset the oxidising effect, the smelter has to maintain a strong blast of air through the furnace and to infuse excess fuel at regular intervals, so that the reducing gas carbon monoxide produced in the furnace dominates over carbon dioxide. The occurrence of poor quality iron in small quantities at the earliest levels of Iron Age sites clearly indicate that the iron smelter acquired the prerequisite knowledge gradually over the years through experimentation (Hegde: 1981:196). During prolonged experimentation in the production of iron, much of iron was lost as slag. When the ironsmith was able to balance these conflicting demands, he was able to produce a pasty semi-fused, spongy mass of iron. The iron thus produced required further treatment to remove the presence of non-metallic inclusions and hammering the red-hot sponge iron on an anvil did this. The metal thus produced was used for producing weapons and implements (Joshi: 1970: 38-9). As indicated earlier, Iron-producing technology was not incipient but fairly advanced when it entered Tamilnadu in c.500 BC. The study of iron industry in ancient Tamilnadu is mainly based on the Archaeological reports (exploration and excavation) of industrial sites especially iron and steel producing centres in Tamilnadu and other places in India. The sites excavated outside Tamilnadu were mainly studied to see the technique that went into the smelting operation in these regions and to have a comparative study of smelting process in ancient India.

As mentioned elsewhere, the rich iron ore deposit in South India played a crucial role for the prolific and extensive occurrence of iron equipment in the megalithic and associated remains all over the peninsular India (Banerjee: 1965:190). Indigenous smelting industry flourished in the districts of North Arcot, South Arcot, Chingleput, Salem, Madurai and Tirunelveli. In the North Arcot district, no less than 86 villages smelted the iron ore and the average number of furnaces was about 200 (Balfour: 1855:19). Iron ores, mined and smelted in the native iron industries were found in more than seventy villages in taluks of Kallakkuruchi and Tiruvannamalai in South Arcot district and the smelters in these places purchased the ore from North Arcot district to augment their requirement. (Balfour: 1855: 35-38). Dr.Voysey mentions about the iron-smelting operation carried out near the Red Hills in Chingleput district with the forge and crudely constructed bellows. In Pudukkottai district the history of iron and steel industry can be traced to a period of over 2000 years (*Manual of Pudukkottai State*: Vol. I: 1938:196).

There were two varieties of iron produced from the Coimbatore ores, one remarkable for its softness and malleability and the other for its steel like hardness (Balfour: 1855:65). Smelting of iron took place on the high road from Tiruchirappalli to Madurai and in more than 25 villages in the taluks of Tirumangalam, Nilakottai, Tadikombu and Ramagiri, but in the year 1855 iron was in production only in the two villages of Sivaganga and Kilavadipallam (Balfour: 1855:120). The furnace used in the Madurai district was made of burnt clay. Its height was 75 cm, diameter at the bottom 37.5 cm and at the top 22 cm. The furnace was placed in a space of about one meter square, enclosed by a very low wall, through an aperture under which a clay tube was introduced and two pairs of native bellows applied to the outside end. From the above description it can be surmised that the furnace used in Madurai region was very small compared to the one used in northern part of Tamilnadu and the smelters chiefly produced wrought iron from the ore and for the conversion of steel, they used the iron chiefly procured from Salem (Balfour; 1855:117-123).

Iron smelting furnaces appear to have been in operation in all the villages adjoining a site of the ore. But the furnaces in use in Salem district were extremely small in size and simple in structure (Balfour: 1855:198-99). Iron smelting operations were found along the foot of the ghauts in the villages in the taluks of Tenkasi, Cheranmadevi, Srivilliputtur, Nanguneri, Sankarankoil, and Bramadesam in Tirunelveli district. In the year 1855, seventy furnaces were in operation in eighteen villages in the six taluks of Tirunelveli district. These furnaces produced crude iron locally called "*Irotimbookute*" or "*Irumburarti*" in Tamil (wrought iron?) and not steel. Steel was manufactured in the village of Vagaikulam. The rest of the centres manufactured only crude iron (Balfour: 1855:216-18).

Indigenous iron and steel industry, that was carried on for more than 2000 years, came to an end towards the close of the last century mainly because of the raise in the price of charcoal and increased facilities in obtaining imported English iron, since the opening of the Railways. Even when the industry was tottering, it showed its intrinsic merit; for it was then (1846) that the famous Ironsmith Arunachala Achari from Salem astonished the whole of India by making his celebrated hunting weapons out of indigenous steel produced in Tamilnadu (Belliappa: 1964: 26)

While the people who carried on the smelting operations in North Arcot district were called Manavalu, such persons were called as Shanar in Chennimalai region in Coimbatore district. The smelters in Venkatagiri area of North Arcot region smelted the ore in the summer months. The furnaces were built in the open air and had a lateral slit for letting out the vitrified matter (slag). The smelters took the mass of iron from the furnace, cut it into two immediately and they sold it to the blacksmith in this condition (Buchanan: 1988: vol. II.285).

Dr. Heyene visited the North Arcot region after Buchanan's survey through the district. In his report, Heyene states that the works or furnaces were under a Banyan tree, near a village Yerragutty about 7 km from Sattur. The smelters were exposed to the inclemency of the weather, without any shelter other than the shade of a tree. The smelters collected the ore bearing fine black sand from the rivulets in the wet season or immediately after the rains and smelted iron in the dry season of the year from the beginning of January to the end of March (Balfour: 1855:14).

Furnaces were made of red loom mixed with sand and consisted of two parts. The lower and the larger was about three spans in height and 30 cm in diameter, quite cylindrical in shape and erected over a hole in the ground, about 10 cm in depth; its sides were uniformly about 5 cm thick. The upper part was conical, with the higher portion of the cone reversed, it was about 45 cm in height and at the opening 30 cm in diameter. The bellows were made of sheepskin, a hole being left at the bottom of the cylindrical part of the furnace to receive the nozzle. The two parts of the furnace was cemented with loam and filled the bottom part of it with charcoal. The ore and charcoal were added alternately and the bellows were constantly plied. This produces sufficient heat, if not to melt iron but at least to soften it and conglutinate it with the dross. At the end of the operation it was found in a solid mass at the bottom of the furnace; water was then thrown upon it, and while hot it was cut into pieces, which however were not entirely separated from one another. In that state it was sometimes sent to the market. The iron thus produced was of very inferior quality being porous and full of dross and charcoal, and so brittle that parts of it may be easily knocked off by a few strokes of the hammer (Balfour: 1855: 15-16).

Voysey however describes the compact texture and the brilliant white colour of the wrought iron indicative of complete cementation of the iron used for conversion into steel (*JORASB*: 1832:253-55). But more commonly it was put a second time into the fire and subject to the action of the hammer. By this process it lost two sevenths of its weight and the iron become perfectly malleable and fit for all purposes (Balfour: 1855:16).

Heath in his paper on iron and steel in India mentions that the smelters at Salem erected a furnace that varied in size from 0.90 m to 1.50 m in height from the surface of the ground and the ground was hollowed out beneath it to a depth that varied from 0.20 m to 0.30 m The Pear shaped furnace measures 0.60 m in diameter at the ground, tapering to about 0.30 m at the top. A semi-circular opening about 0.30 m in height and 0.30 m in diameter was made at the bottom of the furnace and covered with clay. Built entirely of clay, two men could construct a furnace in a few hours. The air was provided by two pair of bellows made of a single goat's skin, with a bamboo nozzle. The furnace was then filled up with charcoal and a lighted coal being introduced before fixing the bellows. The fuel and the ore were put into the furnace without adding any flux and the smelters kept up a tolerably uniform blast by working the bellows alternately for four hours, after which a pair of tongs removed the bloom from the bottom of the furnace. The bloom removed from the furnace was further beaten with a wooden mallet while it was still red hot to remove the vitrified oxide of iron and then cut into two parts to show the quality of the interior of the mass. The process of forging the iron into bars was performed by sinking the blooms in a small charcoal furnace and by repeated heating and hammering, the vitrified and unreduced oxide of iron was removed and a small bar of iron 0.30 m in length and 5 cm in breath made from the bloom (Heath: 1837: 187).

Iron smelting in the Salem and Tiruchirappalli region in 1851 was done in a bowl furnace and is described in detail by Fischer. The bowl furnace in the Salem region according to him was about 120 cm in height and 45 cm in diameter at the bottom and 27 cm at the top and terminating in a flat earthen ware pan about 15 cm deep and 75 cm across, having a hole in the centre corresponding to the top of the furnace; this was fixed on the top as a sort of funnel and to prevent the fuel falling over when heaped up. The bottom of the furnace was sunk about 15 cm below the surface of the ground. The first 45 cm of the furnace, round in plan, was built very solid and forming a platform some 60 to 90 cm broad. The ground plan was completed to about 2/3 of a circle, the remaining segment being left open. This lower part was surmounted by an earthen ware cylindrical chimney, about 60 cm in height, strongly hooped with iron and plastered inside and out with fire clay. The incomplete part of the circle in the lower part of the furnace was the place where the blast was introduced and where the metal was withdrawn and was closed in a temporary manner when the furnace was in use.

To prepare the furnace for work the open segment in front was built up about 15 cm with clay and an earthen ware pipe about 27 to 30 cm in length was placed upon this, inclining downwards at an angle of about 30° so that the mouth of the furnace was within about 20 cm of the bottom of the hearth and nearly in the centre of the furnace. The reminder of the front portion of the furnace was built strongly with clay. The bellows consisted, of two bullock skin sewn up so as to form bags. A nozzle being at the neck portion and two wooden pieces were sewn down to the other end. So when the pieces of wood were held together the bag was well closed and when opened the air will flow inside.The nozzle of the bag thus made was then introduced into the pipe in the front side of the furnace and well luted round with clay, leaving a small hole just over and between the two nozzles through which the interior of the hearth can be seen (Fischer: 1851: 103-107).

The furnace was filled with 3/4 of charcoal and then about 2.75 kg of broken pieces of ore of the size of bean and as this burnt down, the ore and charcoal were introduced alternately and fired for two hours. On the completion of smelting operation the bellows were withdrawn and the front portion of the furnace was broken down and a mass of iron and slag extricated. The red-hot iron was further beaten with heavy wooden hammer to get rid of a portion of the slag and then with axes used as cold chisels the mass was cut into two pieces. In that state, it was sold to the blacksmith (Fischer: 1851: 108). According to Fischer the smelters used to make four drafts per day by re-building the front portion of the furnace with fresh wall and new pipes for bellows and tuyere. The quality of iron varies a little with the quality of charcoal used (Fischer: 1851:109).

Banerjee states that the furnaces in use in South India were circular on plan and conical in shape, broader at the base than at the top. They were about 2 to 4 ft (61to 122 cm tall, about 10 to 15 in. (25 to 38 cm) across at the base and 6 to10 in. (15.25 to 25 cm) at the top. There were two openings at the bottom, one for letting in the air (blast) and the other for extracting the slag. The fuel of charcoal and iron ores was introduced from the top (Banerjee: 1965:186).

2:3: Steel - Developments in Production Techniques

The indigenous furnaces could not develop a temperature more than 1180°C and the carbon absorbed in the smelting process was lower. Because of this the iron produced became wrought iron. The wrought iron thus produced was transformed easily into steel by case hardening here and there. Forge welding of carburised iron plates is evidenced from iron hoe from the excavations at Dhatwa and it is datable to the middle of the first millennium BC (Bharadwaj: 1981-82:142). Steel was produced in India after a few centuries by adopting carburisation process, a process of making homogenous steel, popularly known as wootz. This variety of steel might have been innovated in India around 500 BC and this process could have continued till modern times. Indians innovated steel manufacture by crucible process very early in the history of iron metallurgy and the making of real homogenous steel was a mystery and not known to the Chalybes, Greeks, Parthians and the Chinese (Bharadwaj: 1973: 144). Crucible Steel was produced in India by two processes viz., carburisation of wrought iron and de-carburisation of cast iron (Foote: 1864: *Records of G.S.I. Vol.4*, pt.2., p.377).

Steel was produced at Turkey, Iran, and India from very early times; the technology is similar with regional variations depending on the quality of iron ore and of the availability of fuel as well as on specialised local methods. wootz, the highest quality steel was made in India. The most highly prized were swords of "Damascus steel," originally made in Syria, often from imported Indian steel. The Damascus blade was characterised not only by its functional qualities but also by the etching on the surface of the steel with acid. This revealed an attractive pattern of irregular wavy lines associated with variations in composition of the metal. The Indian wootz steel was the subject of investigation in the

1790s in Sheffield in England, where it was used to make specimen blades of a quality, which could not be replicated by other means. (Arnold Pacey: 1970: 79).

The Indian steel from early times is noted for its strength and cutting qualities. The classical Mediterranean literature mention the making of finest cutlery article and weapons from the steel imported from India. The earliest account of these come from the Greek physician Ctesia of the late fifth century BC, who mentions the present of sword made from finest Indian steel. Similar references to the import of Indian steel were made in Pliny's *Natural History* and *Periplus of the Erythraean Sea* (Banerjee: 1965:160-61). Pre-Islamic Arabic literature and early Islamic writers refer to the swords from Al-Hind or Hinduwani or from India. Tavernier in 1679 mentions the import of wootz steel from Golconda by the merchants from Persia and the making of damascene pattern by the Persian artists by etching it with vitriol (Sarada Srinivasan: 1994: 50-51). Arnold Pacey mentions the fabrication of musket and barrels from the steel used to make Damascus blades by the metallurgists in Turkey from sixteenth century AD. According to Pacey, the steel was first forged into a long strip and then this was coiled into a spiral, using a simple machine with a heavy flywheel to twist the spiral quickly while the red hot metal retained its heat. When the edges of the strip were firmly in contact, they were welded together so that the coil becomes a uniform tube. Finally, the completed barrel was etched with acid to show the damascene pattern running along the gun with the same alignment as the seam (Arnold Pacey: 1970: 81).

The nineteenth century travellers mention regional variations in the making of wootz steel. The three different types of crucible process in use in South India at that time were described in their reports as the Deccani or Hyderabad process, the Mysore process and the Tamilnadu process. The difference between these processes is mainly in the method of cooling the crucible after the smelting operation. While the Mysore process involves the fast cooling of the crucible containing steel ingot by quenching with water, the Tamilnadu smelters allowed the crucible ingot to cool slowly after the operation. The Deccani process involved the fusion of low and high carbon cast iron with a glass piece or conch as flux in the crucible but without adding other carbonaceous matter thereby producing an alloy of intermediate composition (Sarada Srinivasan: 1994: 51-52).

wootz steel was produced in the neighbourhood of Tiruvannamalai in South Arcot district by mixing grey cast iron and malleable iron in a crucible, a process unknown to the steel makers at Salem and other places in Tamilnadu, which closely resemble the Deccani process (Chakrabarti: 1992:145). The manufacture of wootz by Deccani process confined to a few families in the Tiruvannamalai region in Tamilnadu indicate either the migration of artisan community from Hyderabad region or to a close cultural and trade contact between these two places.

The accounts of nineteenth century metallurgists like Buchanan, Pearson, Voysey, Campbell, Heath, and others illustrate the technique of steel manufacture and active production centres in South India. In Tamilnadu, steel was produced in commercial scale in Salem, North Arcot, South Arcot, and Coimbatore districts. Pearson in his paper on wootz steel, in the *Philological Transactions* (Vol.xvii), states that steel produced in India was made directly from the ore by fusion and that it has never been in the state of wrought iron and the variegated appearance of articles manufactured from it was owing to portions of the oxide of iron having escaped metallization when melted up with the rest of the matter. Similar views were expressed by Stoddart and Campbell, But subsequent investigation in the Mysore region by Heyene and in the Salem district by Heath revealed the carbonisation of wrought iron by adding uncharred wood and green leaves in a closed crucible. Voysey however describes the compact texture and brilliant white colour of the iron used for conversion into steel (Balfour: 1855:200). Balfour in his account on the furnace from Malabar states that the furnace differs widely from the still more primitive forges of Salem and Coimbatore. "The egg - shaped furnace was four times bigger than the circular furnaces used in Salem and Coimbatore regions. The height of the furnace from the ground measures 3 m. Native steel for edging ordinary tools was prepared directly from the crude "loupe" or mass as it drawn from the furnace. The lump was never uniform in quality, but presents various degrees of carbonisation. Portions may always be selected more peculiarly adapted to the production of steel, which indeed approximate so nearly to steel itself, that they require only to be beaten out and tempered to form that metal in its primitive state. The manufacture of cast steel as in the case of Salem was not in practice in the Malabar region"(Balfour: 1855:106-7).

Buchanan says that the furnace for producing wootz consisted of a pit with vertically cut round fireplace below the ground level. At a distance of 20 cm from this pit was surrounded by a 1.5 m high wall. Two bellows were connected to the furnace through two tuyeres sunk into the ground and through which air was forced into the furnace. The clay crucibles into which wrought iron pieces were charged along with pieces of wood and fresh green leaves were sealed and placed in the furnace. As many as fifteen crucibles were placed in the furnace and completely covered with charcoal and fired continuously for four hours. The temperature was maintained consistently over 1000°C through out the operation. On cooling, each crucible produced a lump of steel, which was used for producing tools and objects (Joshi: 1970:51).

Heath describes the manufacture of steel in Salem district, in his article on Indian iron and steel, published in the *Madras Journal of Science and Literature* (Vol.viii). He states that the wootz steel was manufactured in the Salem region by packing the bar iron produced in the bloomer furnace in small pieces to enable them to pack close in a crucible, with a tenth part by weight of dried wood and then covered over with one or two green leaves; the mouth of the crucible was then filled up by a handful of tempered clay. The wood used in the operation was acacia auriculata (*karuvela* in Tamil) and the green leaves from the plant asclepias giganta or convolvulus laurifolia (Heath: 1837:186).

The crucibles were formed of red loam mixed with a large quantity of charred husk of rice, which is very refractory and used for one operation (Heath: 1837:187). Sarada Srinivasan states that the inclusion of rice husk in the making of wootz crucible is a distinctive feature of the manufacture of South Indian wootz crucible. According to her the rice husk were added for their high silica and carbon content, making it a highly effective composite refractory material. This will enable the crucible to withstand very high temperature for a long firing cycle and also complete the carburisation of iron charge in a two to four hour cycle of operation (Sarada Srinivasan: 1994: 56).

The peculiarity of the Indian process of steel making, according to Heath, is that, "the Indian process differs from the ordinary English process, from the circumstance of the iron being put into the crucible in the pure state, and without having gone through the process of cementation or conversion into blistered steel". Mushet in the year 1800 discovered the conversion of iron into cast-steel by fusing it in a closed vessel containing carbon. The substances Mushet proposed to use were charcoal dust, pit coal dust, plumbago or any substance containing the carbonaceous matter. Mushet's specification was mainly based on the principle of the Indian process, which adopt the use of dry wood, a substance containing carbonaceous matter. In the year 1825 Mackintosh took out a patent for converting iron into steel by exposing it to the action of carburetted hydrogen gas at a very high temperature in a close vessel. In this process the conversion is completed in a few hours. The Indian process appears to combine the principles of both the above described methods; on elevating the temperature of the crucible containing pure iron, dry wood and green leaves, an abundant evolution of carburetted - hydrogen gas would take place from the vegetable matter, and as its escape would be prevented by the luting at the mouth of the crucible. The carburetted - hydrogen gas would be retained in contact with the iron, which at a high temperature, appears from Mackintosh's process to have a much greater affinity for gaseous than for concrete carbon. The Indian process would greatly shorten the operation and probably at a much lower temperature than were the iron in contact with charcoal powder. Iron is converted into cast-steel by the smelters of India in two hours and a half with the application of heat. The smelters in England would consider the time quite inadequate to produce such an effect. While at Sheffield it requires at least four hours to melt blistered steel in wind furnaces of the best construction, although the crucibles in which the steel is melted, are at a white heat when the metal is put into them, and in the Indian process the crucibles are put into the furnace quite cold. Experience may have taught the Indian smelters different woods and leaves probably contain carburetted hydrogen in very different proportions and he can make iron pass into the state of steel more quickly and with a smaller bulk of particular kinds of vegetable matter than with others"(Heath: 1837:396-97).

D.K. Chakrabarti quoting David Mushet mentions the manufacture of wootz steel by Deccani process in the Tiruvannamalai region. Mushet's accounts indicate that this type of manufacture was confined to few families in the neighbourhood of Tiruvannamalai and was unknown to the common steel makers at Salem, a distance of only 110 km. According to him, the steel makers in the Tiruvannamalai region produced grey cast iron to mix it in the crucibles along with malleable iron (wrought iron?). They made grey cast iron as a substitute for carbonaceous matter in the manufacture of cast steel (wootz). They produced grey cast iron in a small blast furnace, about two and a half metres (2.5 m) in height and tapering from forty-five centimetres (45 cm) at the bottom to twenty-two centimetres at the top (22 cm). The quantity of charcoal required to make grey cast iron in the blast furnace was about four times more than used in the common charcoal pig iron

Pre - Industrial Iron and Steel Industry

furnace. The grey cast iron thus produced was used in a crucible with an equal portion of the malleable iron produced from the bloomer furnace for the production of wootz. The mixing of high and low carbon iron was considered to be more advantageous than the introduction of loose carbonaceous or vegetable matter along with bar iron into the crucible as was practised generally in India (Chakrabarti: 1992:145).

The accounts of Mushet on steel making process in the Tiruvannamalai region is silent on the use of flux in the crucible and in this the description of Voysey, on the native manufacture of steel in South India, mentions the use of conch or glass along with iron in the crucibles, at Konasamudram situated about 20 km south of Godavari by the manufacturers of wootz steel.

2:4: Production of wootz Steel by Crucible Process

on the development of crucible furnace for the manufacture of steel in India, it is said: "A remarkable development in the early 19th century India is melting of steel in crucibles. Bloomer or non-homogenous steel and charcoal were put into crucibles, which were sealed and heated four hours in a hearth with a forced drought. This is an improvement of the African process of heating pieces of iron in a clay envelope and later forging them. The famed Indian steel, known as "wootz", was usually homogenous high carbon steel with a carbon content of 1 to 1.6% and was exported to the west. It seems to be analogous with the Damascus steel, as this probably the town through which it entered the west during the medieval period" (Kuppuram: 1989:314). The excavations at Kodumanal in Tamilnadu from 1985 to 1996 revealed the manufacture of wootz steel by crucible process as early as 300 BC. This clearly repudiates the view cited above on the development of crucible furnace and the manufacture of steel by crucible process in India.

Tamil classical work *Purananuru* (v.13) as mentioned elsewhere, refers to the artefact made of *urukku* i.e., fused metal or steel. This early literary evidence on wootz steel by crucible process is corroborated from the excavation at Kodumanal, which yielded important evidence on steel manufacture by crucible process as early as c.300 BC. Hence it would be worthwhile to describe them in detail here. Fifteen trenches laid 300 m north of iron smelting furnace area yielded two-crucible furnaces, of which one is found in unused condition. These furnaces were found at a depth of 125 cm below the ground, right on the natural soil. More than 12 small furnaces surrounded the main crucible furnace. The main furnace was oval and measured 112 cm north south and 100 cm east west. The furnace had a depth of 40 cm. The furnace wall made of burnt clay had a thickness of 20 cm. The burnt clay had a rectangular hole made in an acute angle. The complete absence of tuyeres in the crucible furnace indicates the use of natural drought for blast. The small circular furnaces surrounding the main furnace had a diameter of 30 cm at the mouth with a small hole or depression at the centre. The small furnaces were connected to the main furnace through burnt clay pipes. A partially broken crucible in vitrified condition was found *in - situ* in one of the small furnace. The bowl shaped broken crucible had a diameter of 0.9 cm at the top with a total thickness of 7 mm at the top and 9 mm at the base. Beside this, many broken crucible pieces were found nearby (Rajan: 1994:95-96).

The nineteenth century accounts clearly describe the use of bellows and tuyere in the manufacture of wootz steel in the Salem region. They do not mention the use of small crucible furnaces surrounding the main one. However the excavation at Kodumanal amply illustrates the use of natural drought and the use of small bowl shaped furnaces as fast cooling zone instead of forced air blast as was used in the 19th century crucible furnaces in the Salem region. Metallurgical analysis of an iron arrowhead datable to ancient period from Kodumanal revealed widmanstatten structure (Pl.22:2). The Widmanstatten structure occurs in steels, which have been rapidly cooled from high temperature (~ 1000°C) but not quenched with water. The Widmanstatten structure is formed by the ejection of ferrite or cementite along certain crystal planes forming a mesh - like arrangement (Tylecote: 1964:316). The absence of martensite and the presence of widmanstatten structure in the arrowhead from Kodumanal showed that it was a fast cooled one but not water quenched. The post-holes found around the furnace and the floor level indicates that there was a super structure over the workshop. The location of the crucible furnace site in the midst of the occupation area unlike the iron smelting furnace, and also the relatively better living condition of the smelters making steel than their iron smelting counterparts points to the flourishing market for this value added product in ancient Kodumanal. The iron and steel industries at Kodumanal played an important role in Trans-regional trade in ancient period .It is clearly revealed by the occurrence of Roman wares, pot sherds bearing Sanskritised inscriptions in Brahmi script and punched marked coins contemporaneous to the industries discovered (Rajan: 1994:97).

3
Metallurgical Studies of Iron Artefacts of Iron Age

3:1: Typology of the Objects from Iron Age Sites in Tamilnadu

Before we embark on the technical study of the Iron objects datable from c.500 BC to c300 AD, a period that combines both Iron age and early historic age in this part of the country, it is necessary to know the typology of the objects taken up for study. The Iron Age habitation and burial complex in South India has brought to light, iron artefacts of diverse forms in large numbers. Though most of the artefacts were intact in shape, they have succumbed to the vagaries of nature and completely transformed into mineral form. Many of the iron objects were either corroded or thickly encrusted with either particles of gravel, or a mass of fibrous black rust. The metal of these could be crushed to powder between the fingers. Mortimer Wheeler, while commenting on the iron objects from South Indian megaliths refer to the better preserved iron tools in the deliberately and carefully closed and sealed megalithic tombs than those left to themselves without plan or design in the other parts of India. The metallurgical analysis of the iron objects from Sanur megalithics by M.K.Ghosh revealed that the artefacts were free from metallic iron and has transformed into ore form due to vagaries of nature defying protective condition of its preservation (Banerjee: 1965:194). Iron artefacts in the collections of Archaeological Survey of India, University of Madras, and Tamil University at Thanjavur formed the main sources of study for this monograph.

We have provided below the type of iron objects retrieved from fourteen iron age-early historic sites in Tamilnadu. Iron objects from the burial (or megalith) and habitational sites are indicated with the suffix M and H respectively.

Sanur, M&H: Megalith-1 yielded nineteen: the spear (similar to type 2), tanged dagger and knife (type 9) and the hook (similar to type 14); 2- Twenty-six: the spear (similar to type 2), the bar (types 4 and 6), the tanged dagger or knife (type 8), the wedge (similar to type 12), the hook (types 15 and 16), the spear or Arrow –head (type 23), and the sickle (type 23); 3- and 4- Respectively three and two, all fragmentary; 5- Seventy-one: the spear (types1 to 3 and 5), the knife or dagger with tapering ends (type 10), the wedge (type 11 to 13), the hook (type 14), the tanged arrowhead (types 17 to 22), the knife (type 24), and the horse – bit (type 26). Sometimes iron objects, mostly arrowheads, were found beneath the sarcophagus. In megalith-5, a group of arrowheads was found dumped inside a vessel in the Black and Red ware. Otherwise, the objects were huddled in all possible directions. Of the types, the horse-bits and flat bar deserves special mention. Horse–bits were obtained from the cairn packing in megalith-2 (dolmenoid cist) and from inside the pit in megalith-5. Even in their mutilated form, they leave no doubt as to their shape and utility.

They were produced by the bending of the side - bar into larger upper and smaller lower loop. The upper and lower parts of the mouth of the animal could be pressed in position. The specimens are unique. The corresponding objects of the Bronze Age in western Europe and of the iron age in Britain mostly have a linked bit and either vertical side-checks with the side–loops or simply side–rings to fasten the strap but not a solid cylindrical transverse mouth- piece and arched upper and lower checks. Horse bit specimens from Guntakkal, Anantapur district and from Adichanallur, Tirunelveli district (now in Madras Museum), have cylindrical mouth–pieces like the Sanur types but do not have the arched fittings. They have either an oblong lower portion or terminal outward loop, at the top just above the portion. The extremities of the mouth –piece have mere vertical side checks with broad triangular side–buckles to fasten to the strap. Two thick flat objects, with a socket-like folding at one end and a bevelled working edge at the other, both from Megalith-2 need some consideration. It has been said: " much iron went not only to make tools, weapons and useful appliances of many kinds, but to furnish the medium of commerce itself that is a currency" (*Ancient India*: 1959:no.15: pp.34-37). M.K.Ghosh, a former Member of the Parliament and a metal enthusiast, analysed a few samples of iron artefacts from the well-preserved megaliths of Sanur, and reported that the composition was free from metallic Iron. As mentioned earlier, it had disappeared in the course of disintegration that had set in, defying the protective condition of its preservation (Banerjee: 1965:194**).**

Perambir, M: The iron artefacts, exposed from the burial complex, numbered sixteen that include iron hatchets (5 nos. light short axe), chisels (5 nos.), a scythe, a small knife with handle. An arrowhead with handle and a portion of an iron sword handle of an iron sword and a sickle.

Kunnattur, M&H: The excavation of megalithic burials at Kunnattur brought to light, iron artefacts consisting of flat celts, spearheads, knives and nails. All these artefacts were found placed either at the bottom of the pit or inside the sarcophagus in megaliths no. 4, 5 and 9. In megalith no. 4, most Iron artefacts consisted of flat celts, spearheads, knives and nails. However two pairs of horse-bits were found about one foot higher in the pit filling; The megalith no. 5 revealed the presence of an iron spike measuring 144.5 cm inside the sarcophagus along with a few vessels of Black and Red ware. The rest of the iron artefacts were found outside the sarcophagus. The iron artefacts found inside the terracota sarcophagus in megalith no. 9, consisted of two short daggers, adzes, a chisel, and a spearhead. Habitation area: The 60 cm habitation deposit of period-1 was contemporaneous to the megalithic burials at Kunnattur. It was characterised by small pits cut into the bedrock. The period-1 at Kunnattur is devoid of iron artefacts (*IAR*: 1957-58:18)

Paiyampalli, M&H: Iron artefacts recovered from megalith-1 consisted of an arrow head and a fragment of a spear head; Eighteen iron arrow heads were found deposited around the sarcophagus. Iron antiquities recovered from the habitation area in the year 1964-65 include knives, sickles, and nails (*IAR*: 1964-65:40). The enormous quantity of iron slag and the artefacts like sickles, spears, chisels, nails and axes were found from the habitation area. A rectangular axe found here measured 27.5 cm in length, 10 cm in breadth and 3 cm in thickness (*IAR*: 1967-68:31)

Appukkallu, H: The excavations at Appukkallu brought to light, seventeen iron artefacts from the trenches Appukkallu I and II. They were the Iron blades (2), Iron knifes (3), iron nails (6), iron arrowheads (3), iron hook (1) and iron slags. Except for the iron slag (APKL-I, layer.no.14) and an iron artefact (APKL-II, layer-6) the rest of the iron artefacts belonged to the early historic (100 BC-300AD) period and the former belonged to Iron Age period at Appukkallu (*IAR*: 1976-77:69).

Mallappadi, H: The iron objects unearthed at Mallappadi include iron nails (2), iron points (2), cylindrical iron rod or iron bar, iron blade (Pl.4:2), Iron arrowhead, iron ball and iron objects (2,unidentified). The metallurgical analysis of the cylindrical iron rod or iron bar indicates that the artefact was made of three iron plates (wrought iron in the bottom, low carbon steel at the top and in the middle) welded together by hot forging process (*IAR*: 1977-78:67).

Adiyamankottai, H: The iron objects from Adiyamankottai were mainly from early historic period (100 BC – 300 AD) and were found in the trenches Adiyamankottai I-A, II, and III. The objects include iron nails, an iron knife and iron blades (*IAR*: 1981-82:73).

Kanchipuram, H: The iron artefacts from Kanchipuram include Iron arrowheads, billhook (Pl.9:6), points, nails and nails with bulbous head (*IAR*: 1971-72:53).

Kodumanal, M&H: The iron objects include tanged and barbed arrowheads, fragmentary rings, chisels, battle-axe, swords (Pl.5:4), daggers, horse stirrup (Pl.5:1), hoe like object, crow-bar, leaf shaped arrowheads (Pl.5:2), sickles, hooks, bangles, iron-beads, iron-bell, ceremonial battle-axe. The length of the swords placed in the burial complex varied from 4 to 5 feet (120 to 150 cm) (Rajan: 1994:13-14).

Vallam, H: A single Iron arrowhead belongs to the last part of period I-A. Most of the other objects are nails. Some of them cannot be identified, as they are much defaced due to heavy encrustation. One object looks like a chisel. The objects recovered from Vallam are: 1.VLM-1. Period-II: Iron wedge; 2.VLM-1. Period I-B: nail; 3.VLM-I. Period-II: nail; 4.VLM-2.Period- I-B: nail; 5.VLM-1. Period-I-B: Iron chisel; 6.VLM-I.Period-II: nail; 7.VLM-I.Period-I: nail; 8.VLM-I.Period-I-B: nail; 9.VLM-2.Period-I: nail; 10.VLM-1.Period-I-A: arrow head; 11.VLM-1.Period-I-B: nail; 12.VLM.I period-II: Chisel; 3.VLM-I.Period I-B: an unidentified iron object (Subbarayulu: 1984:43).

Kambarmedu, H: Most of the Iron objects from Kambarmedu were iron nails. Other than this a highly encrusted iron sickle was the other artefact found from the Iron Age period (*IAR*: 1982-83:68).

Tirukkampuliyur-Alagarai, M&H: The megaliths at Tirukkampuliur were not excavated. The metal artefacts from Tirukampuliyur and Alagarai excavations include copper and iron. While only a few specimens are obtained in copper, a large number of iron objects have been unearthed. Most of the iron artefacts were in a bad state of preservation and were found distributed fairly in all periods of occupation. Some of the copper and iron artefacts of period-1 are closely parallel to similar specimens from the excavations at sites like Brahmagiri, Maski and Sanur, generally assigned to the first half of the first millennium BC.The iron artefacts unearthed from Tirukampuliyur include: Chisel (TKP-I. Period-I); tanged arrowhead (TKP-I.Period-II); nail of round section with knob head (TKP-II.Period-II); Spike of lenticular section tapering towards one end (TKP-I.Period-I); Broken piece of an indeterminate object (TKP-I.Period-I); Hook with a sharpened end (TKP-I. Period-II); Barbed arrowhead (TKP-I.Period-I); Arrowhead (TKP-I.Period I); a ring (TKP-I. Period-II); Bent barring with an oblong section (TKP-I. Period III); Barbar's knife (TKP-I. Period-III); an un-identified object (TKP-V. Period-II); Tanged blade with a circular oblong section (TKP-VII. Period-I); an indeterminate object (TKP-V. Period-III); a cylindrical rod with a thick bulging portion in the middle (TKP-VII. Period-I); Broken part of a Nutcracker (TKP-VI.Period-II); a thin knife blade with a pointed end and sharp edges (TKP-I. Period-I) (Mahalingam: 1970:15).

Kaveripattinam, H: Iron objects associated with megaliths or urn burial were not noticed in any of the early sites like Kilayur, Vanagiri, Manigramam. Most of the Iron objects from Kaveripattinam (also referred as Kaverpum-pattinam or Pumpattinam) excavation were affected by atmospheric and sub-soil water reaction and pulverised and fragmented beyond identity. It is a coastal site, nevertheless, eight iron objects were found from the habitation site at Vellaiyan Irruppu and one from Pallavanesvaram (Pl.5:3) (Soundara Rajan: 1994:123-4).

Adichanallur, M: The ongoing survey at Adichanallur by the geophysicists Dr.S Badrinarayanan and Dr.D.V.Rao from the National Institute of Ocean Technology, Chennai, indicate extensive mining activity at Adichanallur before its conversion into a burial complex in the early historic period. The excavation of burial complex at Adichanallur brought to light, the largest number of iron artefacts unearthed so far in the excavations of megalithic sites in India (Pls.6:1&6:2). It constitutes the single largest antiquity of all the metallic finds from a site. Implements, arms, or lamps in iron are generally found in conjunction with bronzes. All the swords and daggers have either a spike at the hilt or a curved pick-shaped piece of iron, on to which a wooden shaft was fixed. Spears, arrows, and javelins had a hollow tubular end on to which the wooden shaft was fixed. Iron artefacts unearthed from the burial mound contains iron swords (26)(Pl.7:1), iron dagger (27) (Pl.9:1), Trident and Sulam or lances (9) (Pl.7:2), Spears (12), barbed arrow heads (4), javelins (7), sacrificial daggers (2), iron axes (8) (Pl.8:1), iron hoes (13) (Pl.8:2), large iron hangers used for suspending a number of objects, probably a series of the small iron saucer lamps (9), saucer lamps (13) (Pl.9:3), iron beam rods (18), iron chisels (4), iron tripods (3) (Pl.10:1), and miscellaneous objects such as an iron cylindrical handle ring and a reaping hook. There are number of other artefacts of various kinds, in a more or less fragmentary condition. A sword of the concave edge and ribbed kind, 2'1½"(63.25 cm) long by 2¼" (5.6 cm) broad. The rib on the flat surface runs from the hilt to the point. There is a ring at the hilt for attaching the wooden handle; a parallel-sided angular pointed sword with raised line up the centre from hilt to point 02'4" (70 cm) long by a 2" (5 cm) broad; short-sword measuring 1'9½" (53.26 cm) by 1¾" (4.38 cm) at its greatest breadth and from the hilt to near the middle of the length the two edges are concave, with a raised ridge in the centre of the breadth; a sword with parallel sides and angular point. Length 2'4½" (71.5 cm) by 2½"(6.5 cm); a curved edged sword with its greatest breadth at its centre of the length. From there, it tapers convexly to the point, length 2'10½" (86.5 cm) by 2"(5 cm); a dagger with spiked handle and worn angular point, length 9¾" (24.38 cm) by 1¼" (3.13 cm). A dagger with a blade resembling a spear. The blade tapers near the point where it is rounded. The spiked handle measured 3½"(8 cm), while the blade is 5¼" (13 cm) by ¾" (1.88 cm); a dagger 12½" (31.25 cm) by 1½" (3.75 cm). The pointed handle is 2¼" long (5.63 cm). A lance or Sulam consists of a long shaft with a bent cross bar fixed at the top of the shaft. The sizes of the Sulam varied from 3'7½" (98.75 cm) to 1'11½" (58.78 cm) in length and the breath varied from 6½" (16.25 cm) at the base of the crossbar to 8½" (21.25 cm) at the points. The-iron hoes were numerous and were of the same design throughout, though differing in form. The hoes were made of thick metal, and all have around projection on each edge of the butt end. The iron hangers have been used for suspending a number of objects, probably small iron saucer lamps. They have a strong broad suspending ring at the top of a vertical rod of thick square section containing four large hooks of various designs at the bottom. Close to the top suspending ring, a series of four to eight arms or ribs branch around, outwards and inwards; iron beam rods were particularly all of one design or pattern. They resemble beams for weighing scales. The longest is 1' (30 cm) to 7" (17.5 cm) by an inch diameter (2.5 cm) in the centre. There are eighteen of them in full shape and pieces of six others. The beam rods are long round metal with an elongated oval bulging in the middle and a knob at each end. Iron tripods from Adichanallur were formed of a ring of a flat metal ¾" (1.88 cm) broad resting on three legs. The legs partly curved outwards, have a bent rest at the foot and they measure 8" (20 cm) in length and 7" (17.5 cm) in diameter. Two other tripods of similar shape but smaller in size were unearthed from Adichanallur burial complex. An iron cylindrical handle ring measuring 1½" (3.75 cm), 1¼" (3.13 cm), and 15. A reaping hook 7¼" (18 cm) were the other artefacts unearthed from Adichanallur excavation (Rea: 1914:pp.131-139). Table-1 given below provides the details of Iron artefacts and iron slag samples collected, from the excavated sites and analysed metallurgically.

Table-1: List of Excavated Sites from where Samples were Analysed

Site	Period	District	Year
1.Guttur	C 500 BC	Dharmapuri	1983
2.Mallappadi	C 500 BC	Dharmapuri	1977
3.Kodumanal	C 300 BC	Erode	1985-96
4.Perur	C 300 BC	Coimbatore	1964
5.Kanchipuram	C 300 BC	Chingleput	1970-75
6.Kaveripattinam	C 300 BC	Thanjavur	1963-73
7.Adiyamankottai	C 300 BC	Dharmapuri	1980-82
8.Adichanallur	C 200 BC	Tirunelveli	1906-12

3:2: Guttur

The archaeological explorations and excavations at Guttur (12° 25' N78° 15' E) brought to light, the existence of an industrial centre dating back to c.500 BC. The ancient metallurgist at Guttur produced cast iron artefacts like in our modern foundry. The production of iron artefact in ancient Tamil Nadu, by casting, was not accidental but deliberate job is illustrated by the recovery of terracotta ring moulds from the excavation of Iron Age industrial sites (Kannarappalayam and Nattukkalpalayam) in Coimbatore district and the reference in early Tamil classical literature *Kurunthokai* (v.155).

The very small quantity of bluish coloured metal piece provided had to be crushed to powder to carry out the following studies. Firstly XRD analysis was carried out and then the powder was given for chemical analysis and wholly used up. The x-ray diffraction analysis using a Philips Diffractometer with microprocessor controlled diffractometer and automatic print out of 'd' values and relative integrated intensities, indicates strong lines of Cu_2O and moderately intense lines of $CuFeS_2$ together with other impurities. Chemical analysis indicates that the powder contained copper of about 50.3%, SiO_2 about 4.9% and the rest being oxides and sulphides of copper and iron. Based on the restricted information available, it appears that the material analysed was copper-based and perhaps not fully reduced to metallic copper (presence of $CuFeS_2$). Having been buried under the ground for a very long time approximately 2500 years, it has got oxidised to the lower oxide Cu_2O. The presence of $CuFeS_2$ indicates that this was probably the ore itself which was reduced to metal, although apparently not completely successfully. (Pl.7:3) Visual appearance of grey twisted strips shows pores all over, apparently indicative of gas escaping from molten metal. XRD analysis seems to indicate very good agreement with the phase Fe_2SiO_4 (fayalite), no other phase being present otherwise in appreciable amounts. This is likely the ferrous slag produced during melting. There do not appear to be aluminate or other lines of appreciable intensity; the material is apparently-completely single phase, viz., fayalite, indicating that good quality ore (rich Fe_2O_3) was melted using a silica lining. Very likely, it was reduction from the liquid state, as the pores on the surface of the strips would indicate. Chemical analysis indicates the presence of about 0.24% phosphorous and 0.03% sulphur in the sample.

X-ray Analysis of an iron piece from Guttur (Pl.7: 4): This piece was stated to have been obtained from the same location as II. The object was badly encrusted. The diffract metric picture was not very strong probably because of the heavy encrustation. The picture seemed to indicate the combined presence of Fe_2O_3 and Fe_3O_4.

Microstructural Studies: A specimen sectioned from an iron artefact was polished with fine alumina powder after grinding the surface with suitable grades of emery paper. The polished surface was etched with 2% Nital to reveal the microstructure. The microscopic examination across the cross section of the specimen revealed a widely varying structure. At one zone, the microstructure showed dark etching pearlite, white etching iron carbide called cementite and ledeburite. The micro hardness measurements giving values of around 900 VPN in the white region of the structure confirms that the phase is cementite (Fe_3C) as shown in (Pl.12:1). The ledeburite structure is a transformation product obtained at 1140°C upon cooling the molten metal from a higher temperature of 1300°C. The microstructure examined at higher magnification gives a clear picture of ledeburite (Pl.12:2). Primary cementite is also seen (Pl.12:3) in some cases as thin white platelet and secondary cementite, which looks like channels along the prior austentic grain boundaries (Pl.13:1). The matrix structure is fine pearlite. The presence of primary cementite indicates that the carbon content in the metal would be higher than 4.3%. In addition to black etching pearlite (fine), martensitic structure (needles oriented at specific angles to each other) is seen (Pl.13:2) which indicates that a ferrous metal (probably cast iron) was melted (a temperature in the region of 1100-1200°C was perhaps attained in a small volume) and then rapidly cooled such as by water quenching. The average hardness of the material lying around 600 VHN would tend to support such a view point. Appearance of many cracks shows that the material has become brittle due to the hard phases like primary cementite and martensite (Pl.13:3). In making these statements, it should however be emphasised that there was difference between the many small samples among themselves. In some cases, the overall hardness was somewhat lower. Again, in another case, what looked like a bit of slag was noticed giving a hardness value of about 420 VHN. Here and there small reddish particles were noticed occasionally giving a hardness value of about 45 VHN. Could these be copper particles?

Chemical Analysis: Chemical analysis of Iron artefact gave Fe_2O_3 of about 74.2%, Phosphorous 0.16%, and sulphur about 0.08%. The rest is probably largely made up of Fe_2O_3. The appearance (reddish relatively in the outer regions) and brownish black near the core would support the view point that a ferrous object produced has corroded (2500 years?), but since it was buried underneath the surface of the soil, there was not much accessibility to oxygen to the core portions which therefore got converted to the lower form of oxide Fe_3O_4. The peripherals however were oxidised to Fe_2O_3. In any case corrosion has been thorough.

Hardness Studies: Vicker's hardness measurements were made on sample no III with a low load hardness tester using a load of 1kg and also a micro hardness tester with a load of 100g. Hardness values were determined for different regions of the sample by making number of indentations. The results are given in table 2 and 3.

Table-2: Hardness Value of various Iron Pieces in various regions

Sl. No	Vicker's Hardness value (VHN)	Sl. No	VHN
1	619	10	330
2	197	11	239
3	689	12	330
4	591	13	413
5	330	14	182
6	269	15	660
7	689	16	720
8	390	17	515
9	421	18	348
		19	742
		20	276

Table-3: Micro Hardness Value of Iron Piece at Various Regions

Sl no	Vicker's Hardness value (VHN)	S.No	Vicker's Hardness value (VHN)
1	488	6	315
2	245	7	710
3	772	8	455
4	803	9	197
5	206	10	253

Metallographic and hardness measurements threw some more light on the artefact under study especially from the micro structural point of view. Micrographs shown in Pl. 12:1 reveal the presence of ledeburite (confirmed also by a micro hardness measurement giving values of 870-900 VHN in the white etching region).

Manufacturing Process: Assuming that if samples II and III belong together, the overall scenario would probably be as follows. Fairly pure ore was smelted with charcoal in a small furnace lined with SiO_2 giving a fairly pure slag of fayalite, Fe_2SiO_4 that was removed. The metal, very probably cast iron, was perhaps water quenched to give the structure of ledeburite, primary cementite, pearlite, and martensite. The presence of some pearlite would indicate that the quenching was not very effective all over. That the whole thing was not a solid-state reduction appears to be revealed by the presence of pores in the twisted strips (constituting Fayalite). From the photographs of (parts of) the furnace (Pl.1:3) provided, it appears that it was perhaps a twin-furnace with a single entry for air (provided with a bellows) in the middle. After putting the ore and charcoal in the silica-lined furnace, it was closed and sealed at the top. After melting, the slag floating at the top was baled out and the metal cooled by pouring water on it. Vicker's hardness measurements were made on sample no. III with a low load hardness tester using a load of 1 kg and also a micro hardness tester with a load of 100 g. Hardness values were determined for different regions of the sample by making number of indentations. The results are given in table 2 and 3.

3:3: Mallappadi

Archaeological Study: The village Mallappadi is situated 2 km southeast of Barugur in Krishnagiri taluk of Dharmapuri district. The habitation site (Pl.11:2) 12° 31' N & 78° 15' E is located exactly on the other side of the hill opposite to Paiyampalli, Neolithic and Iron Age Habitation - cum - burial site mentioned already. It is about 5 km from the latter site. The excavations in the year 1977-78 by the Department of Archaeology, University of Madras brought to light, three cultural periods (IAR: 1977-78:67) similar to Paiyampalli except that the Neolithic culture witnessed a separate entity at the latter site (IAR: 1967-68:31). The megalithic culture is found in association with the lingering vestiges of Neolithic elements in the lowest levels of habitation at Mallappadi.

The occurrence of iron slag and iron artefact at the lowest level of period I in the trench Mallappadi I (MPD.I) testifies the production of iron from the time of occupation by the Iron Age people c.500 BC. The metallurgical skill of the early Iron Age settlers at Mallappadi is evidenced by the analysis of a iron artefact from MPD.I. The metallurgical studies of iron-bar (cut into two pieces for metallurgical study) and rusted iron artefact reveal that the early metallurgists at Mallappadi not only produced wrought iron and carburised them to steel but also hot forged low carbon steel with wrought iron to get strength.

Microstructural Examination of Iron piece -- The longitudinal Cross Section (side view): The Macrograph for the iron piece found in the Mallappadi excavation (Pl.10:2) shows similarities with the currency bar found at excavation at Sanur Iron Age habitation-cum-burial site in Chingleput district excavation (*Ancient India*: no.13: 1959:p.35). The microstructure in the longitudinal section revealed a coarse grained structure with a network of ferrite. The morphology of ferrite and pearlite is similar to widmanstatten structure (Pl.14:1). Some regions of the microstructure revealed streaks of sulphide inclusions, which look like manganese sulphite (Pl.14:2). There are zones in the microstructure where pearlitic grains are found segregated (Pl.15:1). At higher magnification the pearlitic structure is not resolved into lamellar structure, however fine particles of carbides can be observed in the pearlitic structure (Pl.15:2). Pearlitic structure is noticed in almost entire region of the micrograph indicating high carbon area in the steel. The micro structural study reveals corrosions in the iron bar and the brownish corrosion product is observed in the longitudinal view of the iron bar (Pl.16:1).

Fig.2: Top View of Longitudinal Cross of Forge –Welded Iron Bar Section
(Legends: A=Equiaxed; B=Widmanstatten; C=Equated)

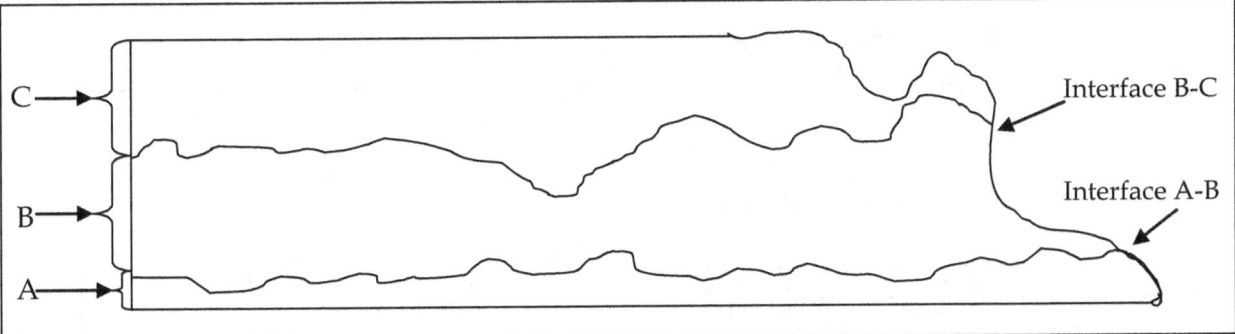

Micro structural scanning of the top surface of the iron-bar clearly revealed that the iron-bar was made up of three layers of metals (Fig.2) joined together by forge welding such as hammering of hot metals. The thickness of the metal A, B, C respectively consists of 0.8 mm, 3.4 mm, and 4.5 mm. The microstructure in the longitudinal surface (top) shows the interface between the two metals at the centre region (Pl.16: 2). A thin layer of slag is found at the interface of the materials (B and C) (Pl.17: 1). The structure in the interface region has cracked probably due to metallographic grinding and polishing of the artefact. The widmanstatten like structure containing mostly ferrite and few plates of pearlite is found in the interface region B (Pl.17:2). The other side of the interface (layer C) showed equi-axed grains of ferrite with no pearlite. The interface containing equi-axed grains of ferrite is shown in (Pl.18:1). Iron carbide (cementite) is distributed throughout the matrix as thin needles. (Pl.18:2). Another layer of metal (Layer A) is present in one side of the iron-bar. The structure of this layer consists of equi-axed grains of ferrite with small grains of pearlite (Pl.19:1). Streaks of sulphide inclusions were also noticed in the microstructure.

Transverse Cross Section: The Microstructure in transverse cross section of the iron bar clearly showed three regions in the structure with a demarcation line separating the three as shown in (Pl.19:2). This indicates that three metals were forge welded. The outer part of the forge-welded metal appears to be pure iron with slag inclusions (Pl.20:1). Forge welding normally carried out by heating two pieces of metal at a high temperature of about 800-1000°C and hammering them. At this temperature, the metal becomes more plastic and ductile and therefore two metals can be joined together under the force of hammering. The joining of two pieces of iron under hot forging is called forge welding. The detailed examination of the artefact revealed that the structure also contained slag inclusions (Pl.20:1). The microstructure in another region revealed not much of pearlite but mainly ferritic grains shown in the micrograph (Pl.20:2). The EDAX studies of the iron-bar indicate that the metal is mainly made up of iron with a small percentage of copper. The EDAX analysis of the regions containing slag inclusions revealed that the slag mainly consists of iron, calcium, silicon, potassium and copper. The slags are normally made up of compounds such as oxides and silicates of iron, calcium, etc.

EDAX Analysis: Table shows elemental composition in different regions of the transverse cross section of the sample. The slag elements in the iron material are revealed by the presence of silicon, potassium, and calcium.

Table – 4: EDAX Analysis and Graph of Iron Bar

INTE-%; LABEL = Needle Ppt of A67/5; 107.261 Live Seconds

ELEM	CPS	WT% ELEM
FE K	1402.155	99.663
CU K	4.214	0.337
Total		100.000

ELEM	CPS	AT% ELEM
FE K	1402.155	99.704
CU K	4.214	0.296
Total		100.000

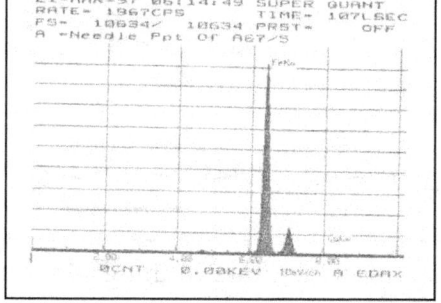

Fig 3.a

ELEM	CPS	AT% ELEM
FE K	353.283	99.821
CU K	0.642	0.179
Total		100.000

ELEM	CPS	WT% ELEM
FE K	353.293	99.796
CU K	0.642	0.204
Total		100.000

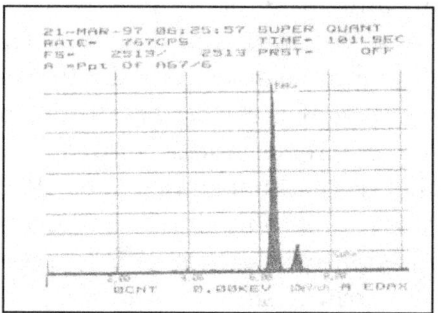

Fig 3.b

ELEM	CPS	WT% ELEM
FE K	441.103	99.888
CU K	0.441	0.112
Total		100.000

Fig 3.c

ELEM	CPS	AT% ELEM
FE K	441.103	99.901
CU K	0.441	0.089
Total		100.000

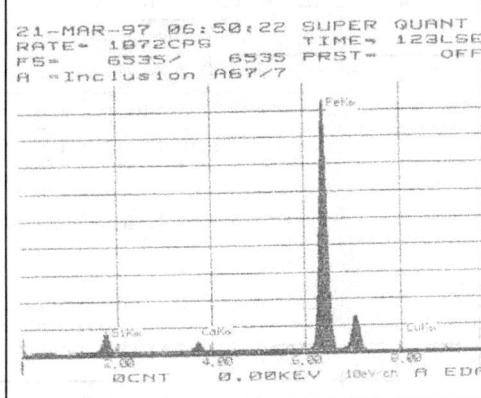

ELEM	CPS	AT% ELEM
SI K	50.396	8.919
CA K	23.837	3.357
FE K	777.682	87.686
CU K	0.342	0.038
Total		100.000

Fig 3.d

ELEM	CPS	WT% ELEM
SI K	50.396	4.740
CA K	23.837	2.545
FE K	777.682	92.668
CU K	0.342	0.046
Total		100.000

ELEM	CPS	WT% ELEM
SI K	86.145	12.315
K K	10.317	1.706
CA K	78.902	12.809
FE K	403.067	72.999
CU K	0.836	0.170
Total		100.000

Fig 3.e

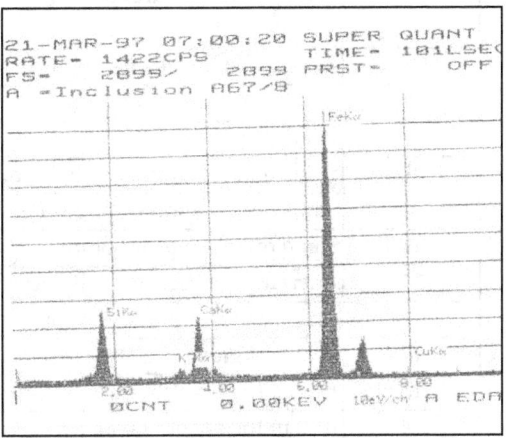

ELEM	CPS	AT% ELEM
SI K	86.145	20.766
K K	10.317	2.067
CA K	78.902	15.135
FE K	403.087	61.905
CU K	0.836	0.127
Total		100.000

Chemical Analysis of the Side Surface (B) of the Iron Bar: The chemical analysis of the iron bar from Mallappadi indicates the presence of other elements such as Carbon(C) 0.9388%, Silicon (Si) 0.1143%, Manganese (Mn) 0.0009%, Sulphur (S) 0.0230%, Chromium (Cr) 0.0008%, Molybdenum (Mo) 0200%, Nickel (Ni) 0.020% Copper (Cu) 0.0131%, Vanadium (V) 0.0041%, Cobalt (Co) 0.0076%

Hardness Studies

Table-5: Hardness Value - Transverse Cross Section

Sl.no	Transverse cross section of the iron bar Area of indentation	Vicker's Hardness Value (VHN)
1	Near to ferritic grains close to crack	118
2	Same region, another point	116
3	Same region another point	114
4	Pearlitic grain structure near grain crack	125
5	Same region, another point	127
6	Same region, another point	128.6
	Longitudinal cross section side view	
7	Ferrite with fine needles	87
8	Same region, another point	93.5
9	Same region, another point	90.5

Table-6: Hardness Value of the Iron Bar, Longitudinal Cross Section, Top View

Sl. No	Longitudinal cross section of the iron-bar. Top view Area of indentation	Vicker's Hardness Value (VHN)
	Layer-A of the Metal	
1	Ferrite region	95.6
2	Same region another point	89.1
3	Same region another point	90
4	Widmanstatten structure and ferrite with some flakes of pearlite.	92.7
5	Another region, same point flakes of pearlite	94.6
	Layer-B of the Metal	
7	Pearlitic region	147
8	Same region, another point Pearlitic region	146
9	Same region another point, Pearlitic region	145
10	Same region another point	140
11	Interface between layer B and C	98.5
	Layer-C of the Metal	
12	Ferrite region	81.8
13	Ferrite region with needles	81
14	It as in the previous no.	82.5
15	Widmanstatten structure in interface between B and C	89.1
	Oxide region in the metal	
16	Oxide region	245
17	Oxide region	237
18	Oxide region	126

The Vicker's hardness measurements were taken on the three layers of the iron-bar with a load of 2 kg. The hardness values varied from 80 to 85 VHN in layer C, 140 to147

VHN in layer B and 90 to 95.6 VHN in layer A. The ferritic structure in layer C exhibited a hardness measurement of about 85 VHN. The widmanstatten structure containing large amount of pearlite in layer B revealed a hardness value of 145 VHN. The layer A of the metal containing equi-axed grains of ferrite and pearlite had a hardness value of about 100VHN. The iron bar is embedded with layers of iron oxide and slags. The hardness of the iron oxide is much harder mainly in the region of 300 VHN.

Manufacturing Process: Solid phase welding is the first welding process used by man without the presence of any vapour or liquid phase in the history of welding technology. The welding of wrought iron has been carried out well above the free running temperature of the slag it contains and the presence of non-solid phase in the metal helps the process (Tylecote: 1968:1). The welding of iron dates from the beginning of the Iron Age because it is implicit in the smithing and working of iron by the direct process. In the early Iron Age the need to weld at an early stage of fabrication was not well understood, and the small pieces of smelted sponge iron were first forged and then joined (Tylecote: 1968:3). This method of welding would be known as smith welding or pressure welding to distinguish it from the large group of welding process based upon fusion. If the joining had been done before forging the sponge into a dense piece, the welds would not have been so visible. This point was seen appreciated since later periods of the Iron Age do not show such a high proportion of badly welded artefacts. The blacksmith welding of iron requires a high temperature of more than 1000°C. The presence of slag in the metal acts as a flux, because its ductility and freedom from oxide films. This helps the metal to weld at room temperature (Tylecote: 1962:152). Iron pieces like adze or axe from Sardis dated to c.1000 BC and Al Mina dated to c.400 BC indicate in the microstructure, a clear seam separating a thin carburised layer from non-carburised region. The joining of thin pieces of strips with higher carbon content (Pearlitic) on to the edges of the axe or adze through hammer welding resulted in Ferritic matrix in the mid region (Maddin: 1992:17-18). Hadfield's micro structural analysis of iron nail datable to Iron Age period from Sri Lanka in the early part of the nineteenth century revealed forge welding. The welding in the iron nail was seen running diagonally across the section, and along the edges of the weld, there are carburised areas (Neogi: 1914:44).

Metallographic Studies of iron hoe from Dhatwa indicate that it was made in two stages. First the red-hot bloom was forged into thin sheets on an anvil near an open-hearth. In this process the surface of the sheets was carburised and casehardened. Secondly, several of these sheets were joined laterally, one by one, by forge welding. Finally, the whole mass was further forged to shape it into a hoe. The laminated structure in the metal indicates the lines of lateral welding (Hegde: 1973:404). Metallurgical study of the iron-bar shows that the iron was manufactured by forge welding of three iron pieces. The inner material is found to be made up of low carbon steel. The layers C and A of the material is made up of pure iron or wrought iron containing considerable amount of slag inclusions. The wrought iron portion on the outer surface being the purest form of iron, has got excellent corrosion resistance and therefore protects the inner mild steel metal from getting oxidisation, even though the metal was made around c.500 BC and buried under the soil for a long period of 2500 years. The inner low carbon steel provides greater strength to the iron-bar.

Streaks of sulphide inclusions were also noticed on all the three layers. Elongation of the sulphide inclusions in the longitudinal direction of the iron- bar indicates that the hammer forging was carried out perpendicular to the longitudinal direction, so that the sulphide slag has spread out in the longitudinal direction. As the forging operation appears to have been carried out at a higher temperature, viscous slag has solidified and embedded between the layers of the metal as well as the surface of the metal. The presence of slag layer on the surface layer was infact quite beneficial to the metal surface because of the protection it has given to the metal from corrosion for more than 2500 years.

3:4: Kodumanal

Archaeological Study: Kodumanal (11° 6' 42" N 77° 30' 51" E) was an ancient industrial and trade centre dating back to c.3rd century BC. The classical Tamil literature *Patirruppattu* refers to the export of industrial goods from Kodumanal to the Roman world. The excavations at Kodumanal brought to light, vestiges of iron, steel, gem stone cutting, polishing and bead making industrial complex, besides large number of metal artefacts (iron and copper) and semi precious stone objects. The iron and crucible steel furnaces and the iron and steel artefacts recovered from the habitation and burial area attest to the thriving industrial and commercial activity at Kodumanal in the ancient period.

Microstructural Examination: Micro structural studies are carried out on the top surface of the barbed Arrowhead (Pl.7: 6). Microstructures at different regions of the Arrowhead were examined starting from the handle region up to the end of the tip. The microstructure near the handle revealed a structure consisting mostly of ferritic grains with few small grains of pearlitic structure. This type of structure corresponds to steel with low carbon content. (Pls.21: 1, 21:2, 22:1) The grain size of the pearlitic structure is found to be very fine. As the structure is scanned towards the tip; the microstructure changes dramatically to a high carbon structure. At a distance of about 41 mm from the handle, the structure consisted of a network of ferrite in a pearlitic matrix. (Pl.22:2). Typical widmanstatten pattern can be observed in the high carbon region indicating that the artefact was fast cooled by air. The grain size has become very coarse in this structure and elongated sulphide inclusions were noticed in this structure. Examinations at a higher magnifications revealed that the pearlitic structure was resolved into a lamellar structure containing alternate lamellae of ferrite and cementite (Pl.23:1). The microstructures containing darkly etched pearlite surrounded by lightly etched cementite suggests the macro-structures associated with the beautiful patterns formed on Damascus steel (Pl.23:2). These patterns consist of well-formed lamellar darkly etched high carbon pearlite with a network of lightly etched iron carbide or cementite; formed by the forging of a high carbon iron ingot followed by etching (Sarada Srinivasan: 1994:56). A large number of corrosion pits were observed (Pl.24:1). Entrapments of slag in large number of packets were noticed both in the handle region as well as in the regions towards the tip (Pl.24:2).

Fig.4.Socketed and Barbed Arrowhead - Different Microstructural Features in Various Locations

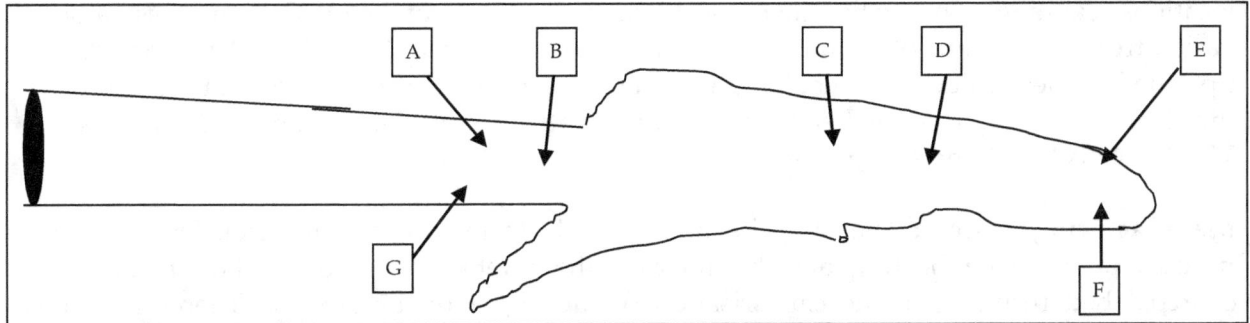

(Legends: A. Ferrite; B. Interface between low and high carbon regions, C. Fine Pearlite Structure; D. Widmanstatten) E ---Fine Pearlitic Structure 0.9% F – Widmanstatten structure, G. - Pearlitic structure)

Hardness Studies

Table-7: Socketed and Barbed Arrowhead - Micro Hardness Values in different regions

Sl No:	Micro Hardness of Barbed Arrow head Area of indentation	Vicker's Hardness Value (VHN)
1	23mm from the socket handle	168
2	Same distance another region	168
3	Same distance another region	168
4	25mm from the handle	131.2
5	Same distance another region	139.2
6	Same distance another region	147.1
7	Same distances another region oxide and sulphide incl.	165.2
8	Same distance another region	128.1
9	Same distance another region	131.9
10	Greyish region	366.3
11	Greyish region	420.5
12	50 mm from the handle, widmanstatten structure	193
13	Pearlitic structure	206
14	Pearlitic structure	225.1
15	Pearlitic structure	206
16	15mm from the tip, high carbon area, 0.6%	272.2
17	Same region Ferrite+Pearlite	213.1
18	Same region Ferrite+Pearlite	206
19	20 mm from the tip, bainitic structure	121.9
20	20 mm from the tip, another region, Pearlitic structure	206
21	Same region, Pearlitic structure	220.5
22	20mm from the tip another region, Ferrite	245.2
23	Same area Ferrite region	206
24	15mm from the tip, Sulphide area	219
25	Same region Ferrite formation	245.2
26	10mm from the tip, net work structure	220.5
27	Same region, widmanstatten structure	236.5
28	High carbon region, nearer to the edge 0.8%carbon	254.4
29	Same distance from the edge another area 0.8%Carbon	254.4
30	2mm from the edge,	264.1
31	10mm from the tip, 2mm from the edge, cementite	296.7
32	Same area, another region, cementite+ Pearlite	254.4
33	10mm from the tip nearer to the centre, Greyish area	541.8
34	Same region, another area, Greyish region	463.6
35	Black region, Pearlitic structure	236.5
36	6mm from the tip and nearer to edge,	321.9

Micro-hardness measurements were carried out on the Vicker's scale at various locations scanning from the handle towards the tip of the Arrowhead. The hardness value varied from a low value of about 125 VHN near the handle to as high as 265 VHN near the tip. The hardness values are calculated in table-7. Hardness study was also carried out on the slag deposit of the Arrowhead. The hardness value on the slag region varied from 365 VHN to 460VHN in the slag pockets.

Manufacturing Process: The arrowhead was made from low carbon steel. During the process of alternate heating and hammering the artefact absorbed carbon from the charcoal bed resulting in the carburisation of the object on its surface. Therefore, the microstructure and hardness of the arrowhead displayed high carbon region in arrow body surface.

Arrowhead (Leaf Shaped and Tanged), Micro structural study: A leaf shaped arrowhead found in Kodumanal (Pl.9:4) is metallographically examined. The length of the arrowhead is 66 mm. Micro structural studies were carried out at different locations of the Arrowhead (Fig.5).

Fig.5 Leaf Shaped Arrowhead - Microstructural Details in Various Locations.

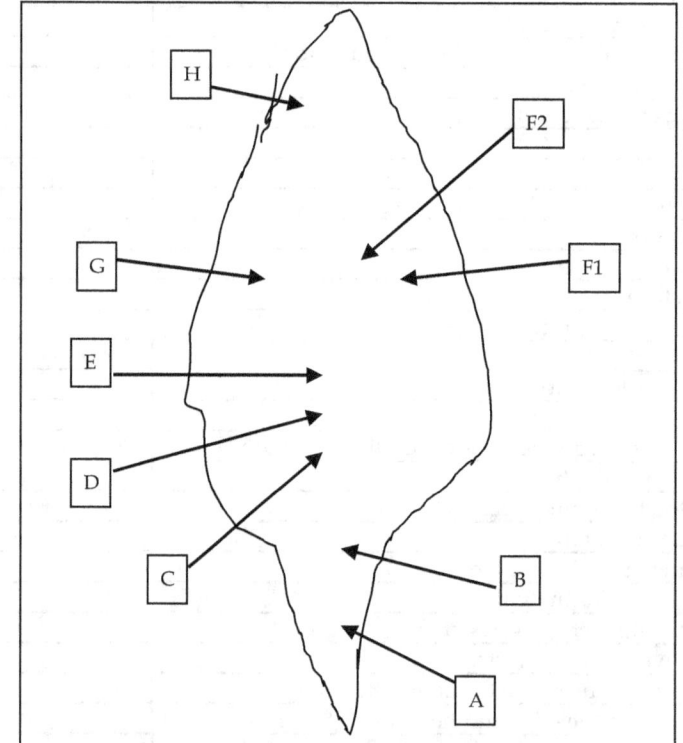

(Legends: A=Slag; B=15mm from the tail end, 5% carbon; C= Pearlitic structure; D=pearlitic region; E=7% carbon, fine pearlite structure; F1=Oxide layer; F2=oxide; G=Slag region; H-Pure iron ferrite region)

The microstructure of the arrowhead consisted mostly of ferritic grains with a few grains of pearlite. The carbon content as estimated from the microstructure appears to be around

0.5%. The composition of carbon in some regions of the arrowhead was around 0.8% to 0.85% and this was indicated by the presence of darkly etched pearlite along with lightly-etched cementite (Pl.25:1).

The structure in the handle region consisted of mostly ferritic grains with a few grains of pearlite (Pl.25:2). The pearlite was not resolved at higher magnification (Pl.26:1). Elongated streak of sulphide inclusions observed in the structure (Pl.26:2). Scanning of the microstructure towards the tip of the arrowhead revealed a change in microstructure. The structure consisted of large amount of Pearlite with a few grains of ferrite (Pl.27:1). The microstructure of the slag region showed silica (Pl.27:2).

Hardness Studies

Table-8: Leaf Shaped Arrow Head - Hardness Values at Different Regions

Sl No:	Micro Hardness of Leaf Shaped Arrowhead. Area of Indentation	Vicker's Hardness Value (VHN)
1	Tail region, nearer to the tang, Slag	441.3
2	Same region, Slag area, another point	513.7
3	Tail region, Low carbon area	128.4
4	Tail region, Low carbon, another area	121.4
5	15 mm from the tail, centre.5%carbon	206
6	16mm from the tail, Pearlitic structure	193
7	20mm from tail, Ferrite structure	160.4
8	20mm from tail, Pearlite structure	213.1
9	Same region another point, pearlite	222.3
10	Same region another point pearlite.	245.2
11	8% C	193
12	Near to the centre, 20mm from tail	724
13	30mm from tail, oxide region	135.5
14	20mm from tail and near to edge 6mm from the tip and 2 to 2.5 mm from the edge, centre portion ferrite region	181.1

It can be noticed that the low steel region in the handle has the lowest hardness of around 125 VHN. The structures of high carbon region in the arrow head had the hardness of 265 VHN. The oxide region revealed very high hardness value that varied from 500 VHN to 700 VHN. The carbon content appears to be around 0.65% in the high carbon region and the pearlitic structure was not resolved even at higher magnification

Manufacturing Process: The steel plate appears to have been hot forged repeatedly to get the required shape and thickness of the arrowhead. The steel plate was hot forged on a bed of charcoal or open hearth at a high temperature and in the process of alternate heating and hammering carburisation has taken place resulting in an increase of carbon content on the surface of the arrowhead. Therefore the microstructure on the surface region consisted of fully pearlitic structure and the hardness value is found very high.

Iron Chisel, Microtructural Examination: The micro structural studies were made on the top surface and the transverse cross section of the iron chisel (Pl.9:5). They showed a structure containing of mostly pearlite (Pl.28:1). In certain regions, a network of ferrite was observed (Pl.28:2). The pearlitic structure was very well resolved at higher magnification. The lamellar pearlitic structure, interwoven in a network of lightly etched cementite (Pl.29:1), as in the barbed Arrowhead examined earlier from the same site, indicates that the iron chisel was produced by forging of high carbon steel at high temperature.

Fig -6: Line Drawing of the Iron Chisel - Microstructural Details at Various Locations

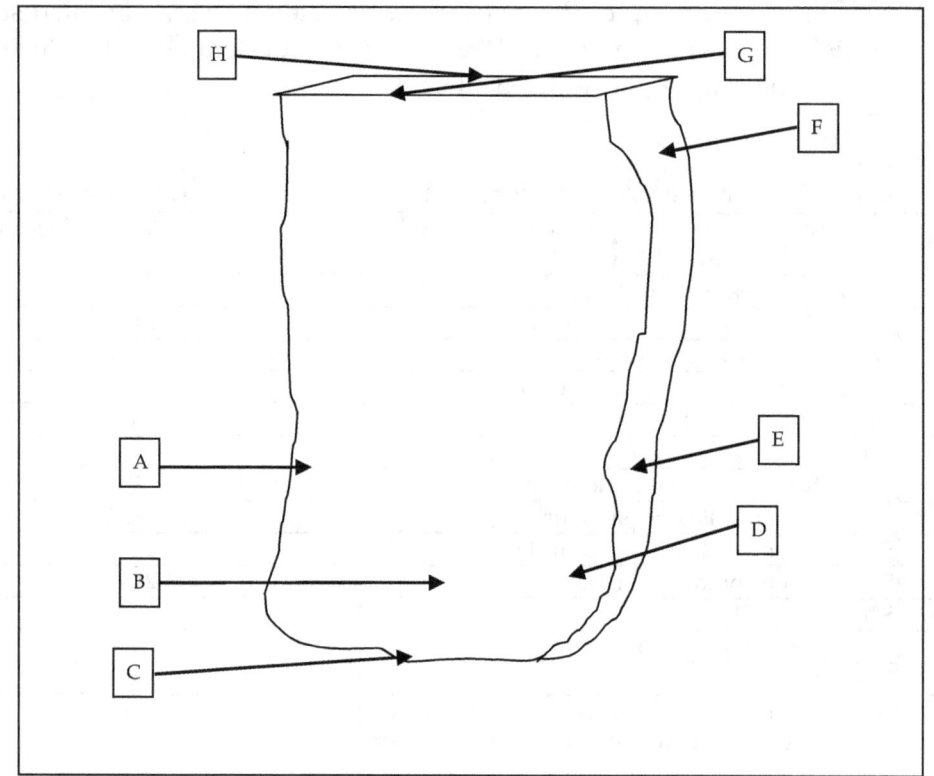

A) Ferrite envelope, B) Widmanstatten structure, C) Pearlite, D) Widmanstatten structure, E) Mixture of pearlite and Ferrite, F) Iron Carbide or Cementite, G. Fine Pearlitic structure, H) Sudden Change in structure from pearlite to Ferrite.

Hardness Studies

Table-9: Iron Chisel - Hardness Values at Different Regions

S. No:	Cross-section of the iron Chisel Area of indentation	Vicker's Hardness Value (VHN)
1	High carbon steel 0.9% carbon	220.0
2	High carbon area Pearlitic structure	193.0
3	Pearite & Ferrite region	181.1
4	Oxide regions	681.5
5	Oxide regions	724.4
6	Oxide regions	350.4

The hardness study by Vicker's hardness measurement indicates that the hardness of the Chisel varies between 180 VHN to 220 VHN. In many regions of the Chisel pockets of oxide and slag was found embedded. The oxide slag contained cubical crystals of silica. The hardness measurement in the slag and other oxide region showed a very high value ranging from 400 VHN to 720 VHN.

Manufacturing Process: The Metallurgical study of the chisel indicates that the chisel is made of high carbon steel. The elongated sulphide inclusions confirm that the chisel was hammer forged at high temperature.

Sword -Bit, Micro Structural Studies: Sword bit from Kodumanal datable to c.300-100 BC was polished and metallographically examined. Though most of the region in the sword bit was oxidised and completely changed into mineral state, the microstructure in one phase clearly revealed ductile iron with graphite nodules (spheroidite). This sword bit (Pl.8:3) was metallographically examined and the study revealed the technology that went into the making of the artefacts. The polished and etched surface of the sword bit revealed a microstructure of ductile iron with graphite nodules (spheroidite) in an envelope of free ferrite (Pl.29:2). Most of the original pearlite has decomposed resulting in a matrix of free ferrite and 5% pearlite (black irregular). The addition of magnesium (or cerium, a modern process unknown in ancient times) spheroidises the graphite, which appears globular in the microstructure. It has to be ascertained by further analysis of iron artefact from Kodumanal that the spheroidal graphite iron obtained was accidental or deliberately made by the artisans of earlier days. Some of the seaweeds contain higher magnesium content (4%) and the artisans probably used these weeds instead of green leaves (*acacia auriculata*) normally used in the crucibles thus resulting in the formation of S.G iron instead of wootz steel. M.D.Kajale (Kajale: 1994:vol.4no.35, 32-33) in his report on plant remains from Kodumanal excavation, refers to the presence of weeds like Euphorbiceae type single weed seed type and indeterminate grass weed, but it is not known about the level of magnesium in these weeds.

Iron Dagger, Micro Structural Studies: A dagger piece measuring 5 cm in length (Pl.9:2) from the excavated site Kodumanal was polished in the centre region and metallographically examined. A small piece from the tip portion of the dagger piece was cut, polished, and mounted on a Lucite mould in order to study the micro structural details in the cutting edge region.

The material of the dagger piece is mainly made up of a low carbon steel (Pl.30:1) with 0.1 to 0.3% carbon.The cutting edge of the dagger has a thickness of 1.0 mm and appears to be forge welded to the centre piece (Pl.30:2). The carbon content of the cutting edge is around 0.8% (Pl.31:1). The high carbon region in the dagger piece consists of darkly etched pearlitic steel interwoven with a network of lightly etched ferrite (Pl.31:2). The microstructure of the centrepiece contained mostly ferritic grains with a few patches of pearlite (Pl.32:1), whereas the cutting edge shows mostly pearlitic grains with globular carbides (Pl.32:2).

Hardness Studies

Table-10: Dagger-Bit, Hardness Values at Different Regions

Sl. No:	SWORD OR DAGGER-BIT Area of indentation	Vicker's Hardness Value (VHN)
1	A. Dagger –bit, cut and polished and mounted tip region Centre portion	160.4
2	Same region	175.6
3	Patches of Ferritic region	193.0
4	Another region, Same as in the previous one	199.3
5	Nearer to the edge, curved portion	170.3
6	Same region, another point	193.0
7	Still closer to edge	186.0
8	Same region another point	165.2
9	Same region, another point	155
10	Near to curvature zone	228.3
11	Same region, another point	220.5
12	Same region, another point	245.2
13	Edge, slag portion	641.7
14	Same region, another point	724.4
15	Centre of the metal, slag portion	605.5
16	B. Dagger -bit polished but uncut big piece region, Pearlitic structure	170.3
17	Pearlitic structure, centre portion, another region	228.3
18	Pearlitic structure, centre portion, another region	206.0
19	Surface coating region (slag)	383.1

The hardness studies confirm that the centrepiece is softer with about 100 VHN and the cutting edge being hard with around 240 VHN. It is interesting to note that the cutting edge is not corroded although it was made 2000 years back. However, the region where the edge is chipped was corroded (Pl.33: 1).

Manufacturing Process: The dagger piece from Kodumanal indicates that the material was made from low carbon steel (0.3%). The cutting edge of the dagger piece was made from high carbon steel (0.8%). The banded structures in the dagger piece indicate that the high carbon cutting edge measuring 1.0 mm in thickness was forge welded to the dagger piece (Pl.30:2). The white coating layer in the cutting edge indicates the use of protective coating (Pl.33:2) and corrosion product can be seen where the layer is broken. *Purananuru* (v.95) refers to the painting of iron artefact with ochre and ghee and it can be surmised that the cutting edge was probably coated with an oxide layer like ochre.

Iron Nail, Metallurgical and Microstructural Studies: An iron nail measuring 2.5 cm in length (Pl.7:5) from the excavated site Kodumanal was polished in the longitudinal direction. The polished and etched surface of the nail revealed the formation of different microstructures over its surface. The micro structural scanning of the surface of the nail revealed different microstructures in various zones of the nail (Fig.7).

Fig –7: Line Drawing of the Iron Nail - Microstructural Details at Various Locations

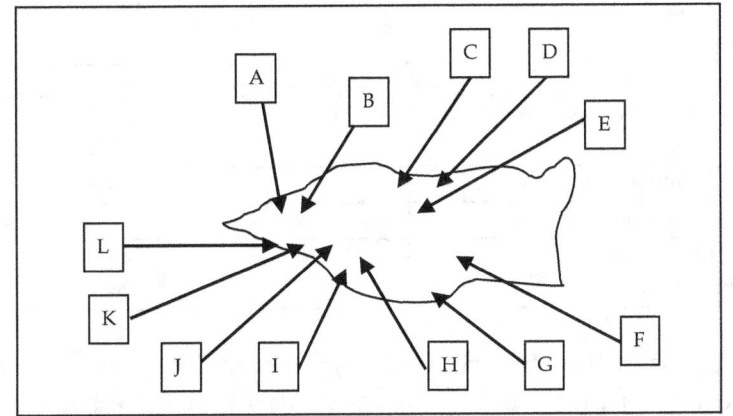

A) Pearlite, B) Slag, C) Slag, D) Ferrite & Pearlite, E) Pearlite, F) Ferrite, G) Sulphide inclusion, H) Ferrite Pearlite, I) Ferrite, J) Ferrite, K) Slag, Sulphide inclusion, L) Ferrite region

About 6 mm away from the tip portion of the nail, the microstructure consisted of single-phase (alpha) grains of ferrite (Pl.34:1). Adjacent to the alpha phase of ferritic structure and at a distance of about 10 mm from the tip portion of the nail, complete pearlite structure was noticed (Pl.34: 2). The presence of pearlitic grain structure indicates that the carbon content in this zone is around 0.8%. The pearlite structure found in this region was very fine lamellar pearlite. The darkly etched fine lamellar pearlitic structure surrounded by lightly etched cementite in the nail piece (Pl.35:1) suggests the formation of beautiful patterns as found on the Damascus sword by the forging of high carbon steel followed by etching. The lengthwise scanning of the microstructure in the middle portion across the section revealed similar type of microstructure containing large amount of pearlite. However the scanning across the width of the nail indicated zones containing structure of wrought Iron revealing equi-axed grains of ferrite and elongated streaks of slags (Pl.35: 2). In many of the regions, oxide and silicate slag layers were embedded in the nail as shown in Pl.36: 1. The structure of slag in Pl.36: 2 reveals solidification of the slag by fast cooling during the forging operation (fabrication of the nail).

Hardness Studies

Table-11: Iron Nail - Hardness values at different regions

Sl.no	Iron nail, longitudinal cross Section Area of indentation	Vicker's Hardness Value (VHN)
1	Centre portion slag +wrought iron region	206
		175.6
2	Wrought iron matrix with iron slag	131.9
		135.5
3	Pure iron. matrix	135.5
		118.9
4	Steel structure sudden change from low hardness to high hardness value	162.5
		160.4

5	Pearlitic zone	199.3
	Pearlite + slag	186.9
6	Pearlite + Ferrite	199.3
	Pearlite + Ferrite	93
7	Pearlitic region+ sulphite sudden change from wrought to steel	
8	Near to the edge, Oxide region	401.2
	Same region another point	383.2
9	Pearlite + sulphite near to the edge	170.3
10	Pearlite structure near to the tip	181
11	Oxide layer near to the tip	383.2
	Slag layer near to the tip	724.4

The micro hardness in the wrought iron region containing ferrite is found to be 130 VHN. The centre portion of the nail showing only Pearlite has the hardness ranging from 170 to 195 VHN. The slag layers are found to be very hard and showing a hardness value of around 725 VHN.

Manufacturing Process: From the metallurgical studies it can be presumed that during the fabrication of the nail, wrought iron piece has been heated in a charcoal furnace (open hearth) and hammer forged. This operation would have been carried out number of times to meet the required shape. While heating the charcoal furnace to a high temperature of around 1000°C, carburisation of the surface of the nail might have occurred due to the carbonaceous atmosphere. The microstructure of the nail therefore revealed pearlitic structure in many regions. Silicate and other slag materials would have become mushy and viscous during the heating and has solidified very fast giving raise to glassy slag entrapment on the surface of the nail.

Iron-Bead, Microstructural Study: A piece of iron-bead unearthed from the habitation area at Kodumanal is by far a truly remarkable artefact, produced by a combined operation of fabrication like metal forming at high temperature and bending and joining by forge welding (Pl.11:1). The side view of the bead is shown in Pl.11:1 and the macro structure of the transverse cross section of the artefact is shown in Pl.10:3. Micro structural examination of the iron bead confirmed the view that the metal was made of wrought iron i.e. very pure iron with streaks of slag inclusions. The microstructure in different regions of the bead revealed equi-axed ferritic grains and elongated streaks of slag inclusions (Pl.37:1). The grain size of the ferrite varies from few microns in the re-crystallised region to 0.25mm in the grain-coarsened region as shown in micrograph of Pl 37:2. The fine re-crystallised grains were found present nearer to the inner surface (Pl.38:1) in some parts of the bead and closer to the outer surface (Pl.38:2) in the other parts of the bead. Therefore, the highly coarsened grains were present very close to the outer surface in some parts and near to the inner surface on the other areas. When the microstructure is scanned from the inner to the outer surface across the section, both the equi-axed fine grains and the coarsened grains were present. The presence of such re-crystallised fine grains indicates that the wrought iron was subjected to cold working and re-heating. Highly coarsened grains in the structure indicate that the growth has occurred after the recrystallization due to exposure to the high temperature for a long duration time.

Hardness Studies

Table-12: Iron Bead - Hardness values at different regions; Load = 100g

Sl. No	Cross section of the iron bead Area of indentation	Vicker's Hardness Value (VHN)
1	Opposite to joint at fine grained structure near to outer surface.	164
2	Same region, another point	138
3	Same region, another point	144
4	Opposite to joint at coarse grained structure near to inner surface.	88
5	Same region, another point	91
6	Same region, another point	89
7	Near to the joint at coarse grained area	99
8	Same region, another point	99.8
9	Same region, another point	106

Micro hardness of the wrought iron-bead showed different readings at different locations. The hardness value in the fine-grained structure regions varied from 164 VHN to 138 VHN. The hardness value in the coarse-grained structure regions varied from 106 VHN to 88 VHN.

Manufacturing Process: The fabrication of the bead therefore has involved production of wrought iron plate by hot forging in the first place and bending of the plate into a cylindrical shape in the subsequent operation. The overall microstructure of the cross section of the bead reveals the slag streaks in elongated oval structure indicating its occurrence in the course of forging and welding operation. During the metal forming operation, temperature of the metal might have varied from a very high temperature of around 800°C to room temperature. Therefore, different sections of the bead have undergone exposure to different temperature for varying length of time. This seems to be the reason for the extremely wide variations in the grain size in different parts of the cross section of the bead. The ends of the plate have been forge welded (Pl.39:1) by hammering at higher temperature. The welded joint of the plates reveals slag and oxide entrapment (Pls.39:2, 40:1) The smooth bending of the slag streaks parallel to the circumference of the bead establishes that the plate has been bent to form cylindrical bead. On the other hand, the ferritic grains have found to be equi-axed and no deformation of the grain structure was seen in the microstructure. The absence of grain deformation and the presence of equi-axed ferric structure indicate that the mechanical working to shape the bead must have been carried out by hot working process i.e. the bending of the iron plate at a relatively higher temperature of around 800°C.

Iron Slag, Microstructural Studies: The microstructure taken from the slag revealed a leaf-like growth pattern (Pl.40:2), i.e., dendritic structure. This indicates that the solidification of the constituents has occurred from liquid phase and the smelters must have attained a temperature of 1150°-1200° C.

X-ray Analysis: X-ray diffraction analysis seems to indicate very good agreement with the phase Fe_2SiO_4 (fayalite) (Fig.8), no other phase being present in appreciable amounts. This is likely the ferrous slag produced during melting. There do not appear to be aluminate or other lines of appreciable intensity - the material is apparently completely single phase, *viz.*, fayalite whch indicates that good quality ore (rich Fe_3O_4) was melted using a silica lining.

Observations: The presence of dendrites in the microstructure of the slag and the presence of fayalite as revealed by the x-ray analysis confirms that the ore used was magnetite.

3.5: Perur

Archaeological Study: The village Perur is situated on the outskirts of Coimbatore City. The excavations at Perur by the Archaeological Survey of India in the year 1970 (IAR: 1970-71:53) brought to light, vestiges of ancient culture datable to c.500 BC. Infact the Black and Red ware pottery of fine thin variety from Perur is comparable to similar pottery types from T.Narasipur in the upper Kaveri valley. This indicates the migration of Iron Age people from the upper Kaveri valley in to the Coimbatore region.

Iron Tang, Microstructural Examinations: The tang of an iron artefact from Perur (Pl.4:5) was metallographically examined. The micrograph showed that the microstructure of iron has undergone complete change into oxides and silicates (ore state) due to its prolonged exposure over the centuries to moisture, air, and earthy matter.

X-ray Analysis: X-ray difractogram studies made on the sample revealed the presence of silica, iron silicate and small amount of pure iron as shown in the difractogram in Fig.9.

Observations: The shape of the iron tang is retained almost completely although the metallic iron appears to have fully transformed into iron silicate (Fe_2SiO_4). It is indeed surprising that no oxides of iron were detected in the tang material. From the thermodynamic point of view, it is possible that the oxides of iron have transformed into silicate of iron due to close contact with silica slag embedded in the iron piece for extremely long periods of time. Hence, it can be concluded that the age of this particular artefact will be considerably more than for the iron artefacts, which have transformed into oxides of iron such as the artefacts excavated from the sites at Kanchipuram. Therefore the relative age of the artefacts can be determined from the state of transformation (oxide or silicate) undergone by the metallic iron.

3:6: Kanchipuram

Archaeological Study: Kanchipuram (12 ° 50'N & 79 ° 75'E) was an important trade industrial centre in ancient Tamilnadu. The excavations at Kanchipuram by the department of Archaeology, University of Madras revealed cultural artefacts datable to c. 2nd-3rd century BC. The carbon –14 determinants of the charcoal samples taken at a depth of 6 metres in KCM1 and KCM 4 near the Kamakshi Amman temple ranges from 2430±135 and 2235±135 years for the period 1A (early historic) in this ancient city.

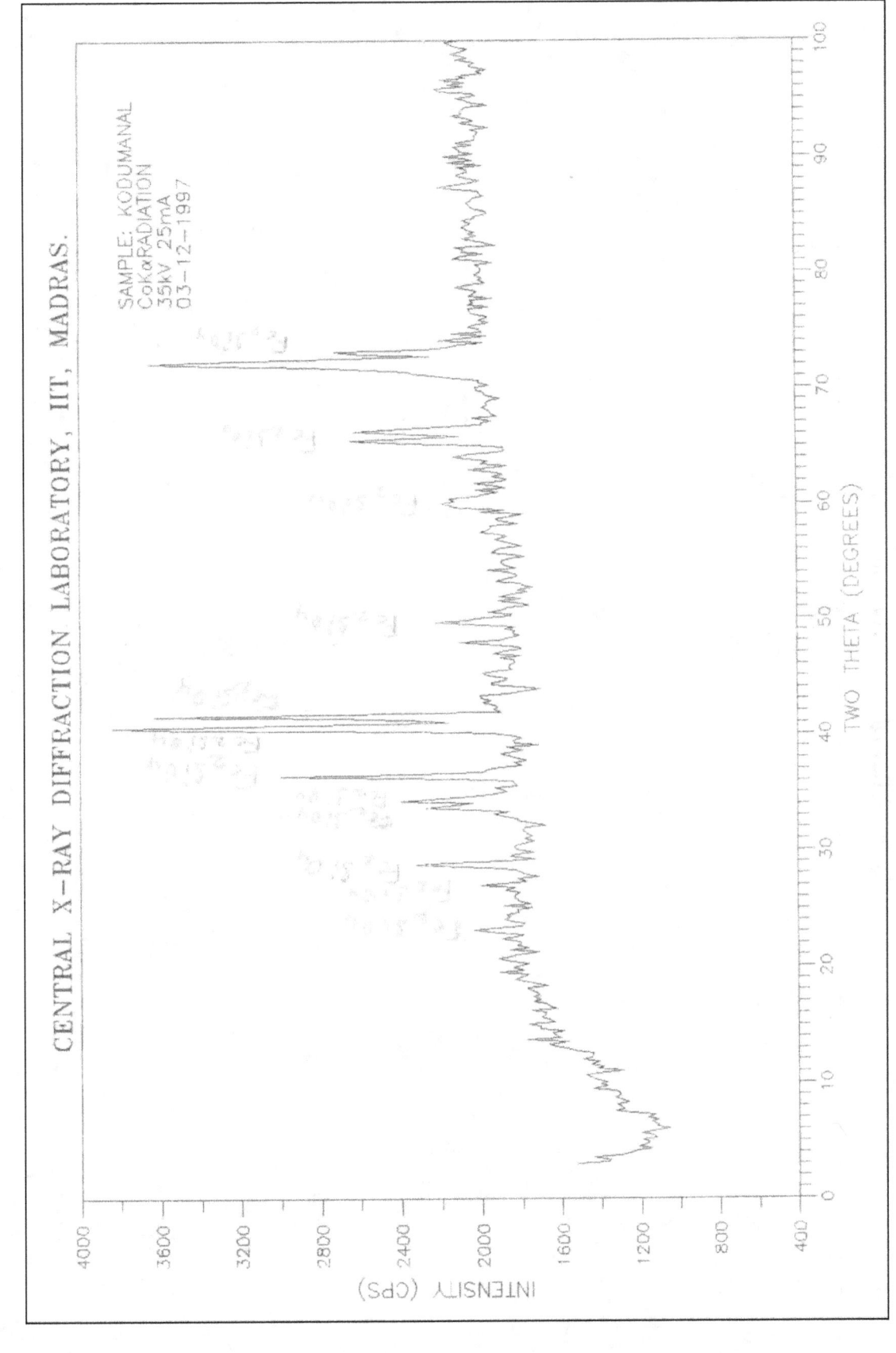

Fig. 8: X-Ray Diffraction of Iron Slag, Kodumanal

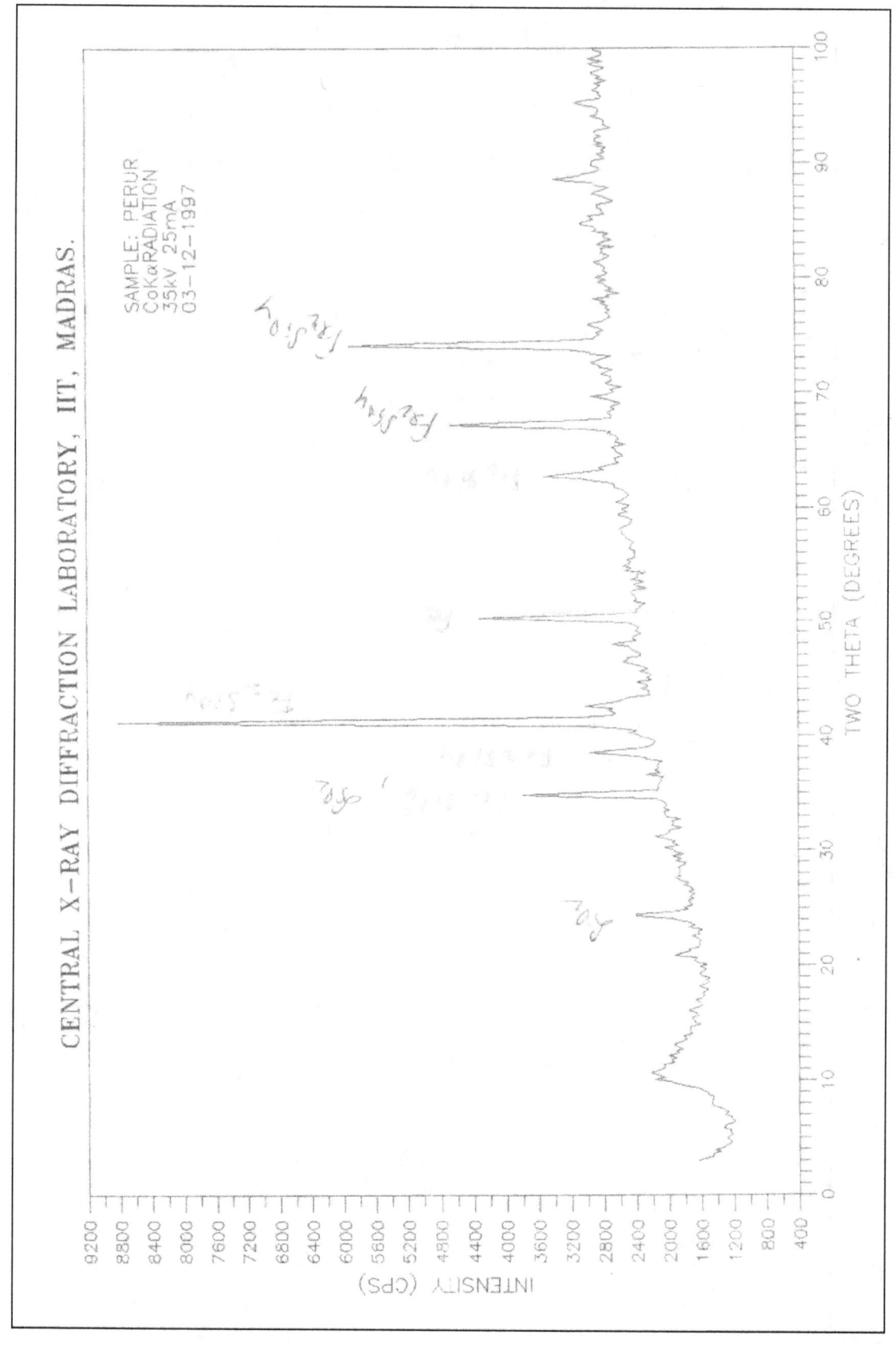

Fig 9: X – Ray Diffraction of Iron Tang, Perur

Microstructural Studies: The iron artefact taken for examination from Kanchipuram was heavily encrusted with soil. To know the exact nature of the iron piece an x-ray was made of the artefact. The x-ray radiograph revealed that the iron piece was bent to form the shape of a hook (Pl.9:6). The polished surface of the iron piece was metallographically examined. The micrograph showed that the microstructure of iron has undergone complete change into oxides (ore state) due to its prolonged exposure over the centuries to moisture, air and earthy matter.

X-ray Analysis: X-ray diffraction studies made on the sample revealed the presence of silica (SiO_2) and iron oxides (Fe_2O_3 and Fe_3O_4 as shown in the difractogram in Fig.10.)

3:7: Mel-Siruvalur
Archaeological Study: The antiquity of the archaeological site Mel-Siruvalur is indicated by the presence of the megalithic cairn circles on the western side of the road in disturbed condition.

Metallurgical Studies: Sarada Srinivasan microscopically examined the lid portion of the iron crucible. According to her, tiny iron prills of a diameter less than 100 µm were found along the glassy edge of the lid. Further analysis of the prill by EPMA confirmed that they are steel prills. The prills were embedded in the outer crucible lining of the lid probably due to splashing of molten liquid on overflowing at high temperatures. The etched microstructure of the prill has a lamellar eutectoid structure of fine pearlite inside hexagonal grains of austentite. Fine pearlitic structure indicates that it derived from a very good quality hypereutectoid high-carbon (>0.8%) steel. The prill had a hardness value of around 400 VPN. The presence of much smaller amounts of a lightly etched network of cementite (iron carbide) and occasional needles in the pearlite was noted between grains near the boundaries (Sarada Srinivasan: 1994:54).

Qualitative analysis of a few samples of the slag collected from the second trench by SEM-EDS (Scanning Electron Microscopy with Energy Dispersive Spectroscopy) showed that the major constituents were iron and silicon, suggesting that these may be fayalite (iron silicate) type iron slags. The composition of the slag indicates that the bloomer process in the trenches smelted the iron charge. The iron bloom produced in the bloomer furnace formed part of the charge to produce high carbon iron by the wootz crucible process in the area where the mound with the crucibles was found (Sarada Srinivasan: 1994:56).

3:8: Explored Sites
Archaeological Study: Iron smelting furnaces and furnace material bearing mounds were discovered in the course of field exploration in the districts of Chingleput, NorthArcot, SouthArcot, and Pudukkottai. The sites where the materials were collected include Kattankulathur, Pakkam, Ramali hills, Veppakkudi, Ponparakottai, Vallathirakottai, and Perungalur. Wet chemical analysis was done on the iron and slag samples collected from the sites Kattankulathur in district Chingleput, Pakkam in district South Arcot, Ramali

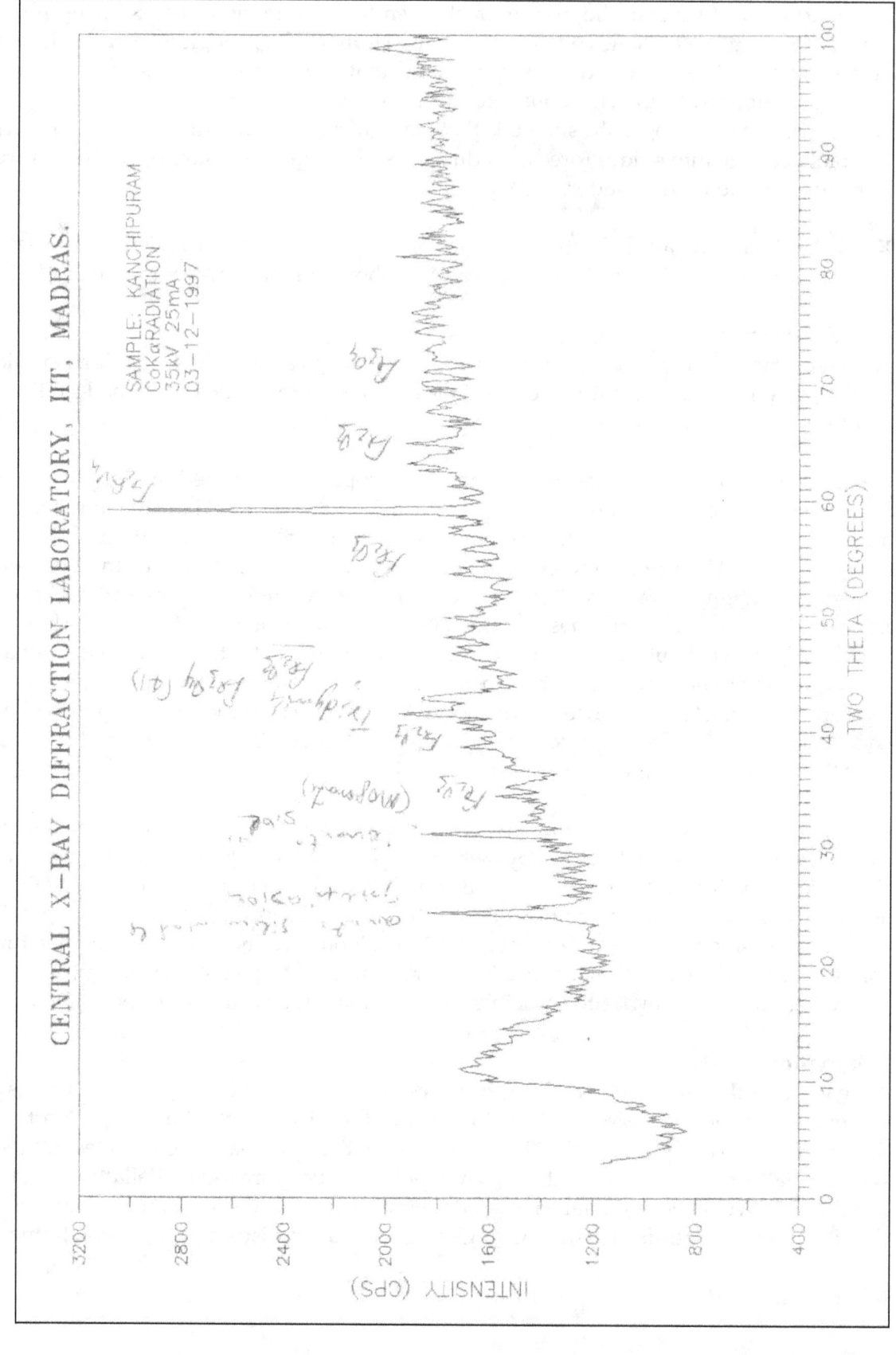

Fig 10: X-Ray Diffraction of Fish Hook, Kanchipuram

hills in district North Arcot, Veppagudi, Ponparakottai, and Perungalur in district Pudukkottai.

Kattankulatur, Wet Chemical Analysis: Wet chemical analysis was done on the iron sample collected from Kattankulathur in Chingleput district. Most of the iron present in the artefact was transformed into mineral form (Fe_3O_4) elemental iron was found in the order of 38.16%. The other elements found in the sample include carbon 2.10%, Manganese 0.58%, Phosphorous 0.038%, Sulphur 0.019%, and Silicon 6.48%.

Ramali Hills, Wet Chemical Analysis: Wet chemical analysis was done on the iron sample collected from Ramali hills in NorthArcot district. Most of the iron present in the artefact was transformed into mineral form (Fe_3O_4) elemental iron was found in the order of 30.10%. The other elements found in the sample include carbon 1.38%, Manganese 0.21%, Phosphorous 0.026%, Sulphur0.010%, and Silicon 6.73%.

Pakkam, Wet Chemical Analysis: wootz steel ingot collected from the surface of the ancient industrial site at the outskirts of the Pakkam village in Sankarapuram taluk in South Arcot district was subjected to wet chemical and x-ray analysis. Wet chemical analysis indicated the transformation of the entire ingot into mineral state Fe_2O_3 due to the exposure to atmospheric conditions. Elemental iron present in the ingot was 1.39%. The other elements traced from the ingot include carbon 1.06%, Manganese 0.10%, Phosphorous 0.016%, and sulphur 0.01% and silicon 6.30%. FeO 19.92% and Fe_2O_3 and Fe_3O_4 29.88%.

Veppangudi, Wet Chemical Analysis: Wet chemical analysis was done on the iron sample collected from the industrial site on the outskirts of the Veppangudi village in Pudukkottai district. Most of the iron present in the artefact was transformed into Mineral form (Fe_3O_4) elemental iron was found in the order of 36.18%. The other elements found in the Sample were including carbon 2.26%, Manganese 0.52%, Phosphorous 0.041%, Sulphur0.016%, and Silicon 7.10%.

Ponparakkottai, Wet Chemical Analysis: Wet chemical analysis was done on the iron sample collected from the huge mound on the outskirts of the Village Ponparakkottai in Pudukkottai district. Most of the iron present in the artefact was transformed into Mineral form (Fe_3O_4); elemental iron was found in the order of 33.23%. The other elements found in the Sample were carbon 1.93%, Manganese 0.39%, Phosphorous 0.026%, Sulphur 0.013%, and Silicon 7.16%.

Perungalur, Wet Chemical Analysis: Molten Iron slag piece was taken from the furnace rim portion and analysed. The study revealed that the major constituent was silicon-di-oxide and its presence on the outer surface was 45.49% and 24.93 % on the inner surface. The other elements present in the slag were (MnO_2) manganese-di-oxide 2.88% on the outer surface and 1.90% on the inner region and carbon 2.30% (C). The presence of iron oxide (Fe_2O_3) on the two regions was 51.63% and 73.13%.

4
Summary

The research on pre-industrial Indian iron was started when Dr. Pearson published his analysis on the famous Indian steel wootz stating that it was made directly from the ore. This kindled the interest of other western metallurgists to study the metallurgical properties and chemical composition of the wootz steel. The accounts of Francis Buchanan and others clearly established that the Indian steel, wootz, was not made directly from the ore but by carburisation of wrought iron with wood and other carbonaceous matters and illustrated in detail the condition of contemporary iron smelting centres, the size of the furnaces and the technique employed by the native smelters of iron and steel.

The beginning of the 20th century heralded the study of iron in its antiquity as against the earlier work on contemporary iron metallurgy. The scholars researched on the origin of iron and its metallurgy from the literary and archaeological data. Schoff, with emphasis on literary data, identified India and the Chera country in the south a probable source that supplied finest steel to the Roman world. Important archaeological finds, from the ancient trade route in the red sea region and at Kodumanal, near Karur, an ancient trade and industrial centre, authenticated the view expressed by Schoff.

Panchanan Neogi was the first researcher to make an integrated approach to the study of iron in India. He made chemical and metallurgical analysis, tensile strength, and the forging and corrosion resistance of early beams and pillars from the archaeological sites. The presence of iron in the painted grey ware cultural level made scholars to assume that the authors of PGW culture are responsible for the introduction of iron in northern region of India and its subsequent diffusion to the other parts of India. The rich iron ores of South India were responsible for the prolific and extensive occurrence of iron objects in the megalithic and associated remains all over South India. Sarada Srinivasan identified three regional variations in the manufacture of wootz steel. They were the Deccani process, Mysore process, and Salem process. Chakrabarti proposed an independent origin of iron in different regions of India as against its diffusion from the northern region and considerable variation and intra-regional differences in the details of iron and steel manufacture. The systematic study on the distribution of iron ores in Tamilnadu had begun around 1800 AD, when Francis Buchanan referred to existence of iron ore in the Baramahal region. He was followed by a number of geologists collecting data on the presence of iron ore in different regions of Tamilnadu. The magnetite - quartzite ore of the Salem - Tiruchirappalli - Arcot region constitute the most valuable group of iron in Tamilnadu.

Summary

The origin and dispersal of iron technology in the Indian sub-continent attracted the attention of scholars in the recent past. The districts bordering Karnataka and Andhra Pradesh formed the nuclear zone of Iron Age culture in Tamilnadu. The excavations at Paiyampalli and Appukkallu and other places in North Arcot district, Guttur, Togarapalli and Mallappadi in Dharmapuri district and Perur in Coimbatore district point to the migration of megalithic people from Karnataka and Andhrapradesh c.6th century BC. The new settlers with their knowledge of mining and metallurgy exploited the rich mineral resources that enriched the pattern of living in the area they settled. The Iron Age in Tamilnadu represents a distinctive phase of culture that came in succession to the primitive neolithic culture.

The Iron Age identified with the authors of megalithic monument in Tamilnadu had a short span of time from c.600 to 200 BC. The presence of metal artefacts especially of iron, besides furnace materials and enormous quantity of iron slag in the lowest stratum of the iron age settlements indicate that a substantial section of the community were artisans by profession. The discovery of twin furnace for making cast iron at Guttur and the iron and steel furnaces at Kodumanal reveal the existence of a class of artisan with professional skill and expertise. The literary sources reveal that the Iron Age society depended on these artisans for articles of superior quality. The occurrence of iron slags and iron artefacts at the lowest stratum of iron age period at Paiyampalli, Mallappadi and Appukkallu in the North Arcot and Dharmapuri districts testify their production from the time of its occupation by the iron age settlers in Tamilnadu.

The metallurgical skill of the early settlers at Mallappadi is evidenced by the analysis of iron artefact from one of the trenches. The metallurgical study revealed that they not only smelted wrought iron and carburised them to steel but also fabricated iron bar by forge welding low carbon steel strip with wrought iron strips to get strength to the artefact.

The excavations at Guttur in the year 1983 revealed the existence of an industrial centre where iron artefacts were produced from c.500 BC. The chemical and metallurgical analyses of iron artefact and iron slag from the exposed twin furnace area indicate that the furnaces were used to make cast iron objects as in the modern iron foundry. Since the excavation is limited in scale, the nature of smelting operation is not clear. We are not sure whether artefacts were made directly from the ore in the twin furnaces or it was done in separate operation like the smelting of ore in a bowl furnace and then using the twin furnace for the manufacture of artefacts. However, the excavations at Kodumanal showed that the smelting of iron from the ore and the carburisation of wrought iron into steel were made in different furnaces and in separate area far away from one another. If this is taken into account then the smelters at Guttur probably produced wrought iron separately in a bowl or pit furnace and the twin furnaces were used to make iron artefacts like iron ring and bells by casting. That the production of metal by casting in ancient Tamilnadu was not accidental but deliberate is revealed by the unearthing of terracotta ring mould from an iron age habitation site in Coimbatore district and from a reference in the Tamil classical literature (*Kurunthokai*, V.155) to making iron bells by casting.

The excavation of burial sites at Sanur and Perambir on Chingleput district brought to light, large number of iron artefacts. The iron artefacts exposed from the burial complex at Kunnattur include iron knives, sword, dagger, spearheads, and two pairs of horse-bits. The artefacts from Kunnattur clearly indicate that majority of them are weapons of war.

The excavations at Uraiyur exposed a cultural artefact contemporaneous to Tirukampuliyur-Alagarai in the lower Kaveri Valley and Kanchipuram in Chingleput district. The presence of a dyeing vat and spindle whorl in large numbers and beads, bangles, iron and copper objects in large numbers in period-I indicate that the early settlers were mostly artisans. The excavations at Adichanallur brought to light, by far the largest metal objects from a single site. Iron technology was not incipient and fairly advanced when it entered Tamilnadu around c.500 BC.

Archaeological excavations at Jodhpura (Rajasthan), Ujjain (Madhya Pradesh), Dhatwa (Gujarat), Naikund (Maharashtra), Guttur and Kodumanal (Tamilnadu) brought to light, vestiges of iron and steel furnaces in ancient India. An iron-smelting furnace datable to c.700 BC was discovered for the first time in the excavations of a megalithic site at Naikund. Iron and steel furnaces were discovered at Guttur and Kodumanal in Tamilnadu in the year 1983 and 1986. The twin elongated oval shaped furnace from Guttur was four times bigger than the furnaces in use in the Salem and Coimbatore region and looks very much similar to the furnaces reported in operation at Malabar by Buchanan. The crucible furnaces at Kodumanal, unlike their 18th century counterparts, used natural drought for the conversion of iron into steel. The smaller crucibles surrounding the main crucible were used as a fast cooling zone where the steel ingots were placed to cool down quickly by air passing through the pipes connected to the main furnace.

Metallurgical studies were made on eighteen iron artefacts and two slag samples from eight excavated sites datable to Iron Age or early historic period and six sites discovered in the course of field exploration. Most of the artefacts were reduced to mineral form due to the close contact with earth for long time; however, metallurgical studies were made possible on eight artefacts one each from Guttur and Mallappadi and six from Kodumanal. The studies showed that the ancient metallurgist in Tamilnadu not only produced wrought iron but also carburised them to steel. They forge welded different kinds of iron to make it rust free and strong. A piece of iron bead unearthed from the habitation area at Kodumanal is by far a truly remarkable artefact produced by a combined operation of fabrication like metal forming at high temperature and bending and joining by forge welding. Based on the analysis of iron artefacts from two sites (Kanchipuram and Perur), Raghunatha Rao of Metallurgical Engineering Department, IIT Madras and this author have proposed a new hypothesis to ascertain the relative age of the artefact on the basis of its transformation from elemental iron into oxides or silicates form. According to the authors, it is possible that the artefact first gets itself oxidised due to corrosion and then transforms into silicate of iron due to its close contact with silica slag embedded in the iron piece for extremely long periods of time. Hence, it can be concluded that the age of the artefact transformed into silicate will be considerably more than the artefact transformed into oxide form. The relative dating of Perur (c.500 BC) and Kanchipuram (c.300BC) sites based on pottery and ^{14}C determinants respectively confirm the above hypothesis since the iron artefacts revealed only iron silicate in Perur and only iron oxide in Kanchipuram. The above study has to be validated by further analysis of artefacts from other sites, and if found correct, it can be useful for relative dating of the artefact, the layer in which it was found and the site in general.

Bibliography

- 1927: on Metals and Metallurgy in Ancient India, *Indian Historical Quarterly*, 3: 121-33, 793- 802 (hereafter IHQ).
- 1929: Iron and Steel in the Rg Vedic Age, *IHQ*, 5: 432-40.
- 1956: The Megalithic Problem of Chingleput in the Light of Recent Exploration, *A.I.*, 12: 21-34.
- Agrawal, D. P. and Ghosh, A., 1973: *Radiocarbon and Indian Archaeology*, Tata Institute of Fundamental Research, Bombay.
- Agrawal, O.P., 1983: Scientific and Technological Examination of Some Objects from Atranjikhera, In: Gaur, R.C., *Excavations at Atranjikhera*, Motilal Banarasidas, Delhi.
- Agrawal, R.C., and Vijaykumar, 1976:The Problem of P.G.W and Iron in Northeastern Rajasthan, In: Gupta, S.P. and Ramachandran, K.S. (Eds.), *Mahabharatha: Myth and Reality*, Agam Prakashan, Delhi.
- Agrawal, R.C., 1967: *Indian Archaeology –A Review*, Archaeological Survey of India (hereafter ASI), New Delhi.
- Atchison, C., 1960: *A History of Metals*, Macdonald's and Evans, London.
- Balfour, E., 1855: *Manufacture of Iron and Steel and the Coals of the Madras Presidency*, Report on the Government Central Museum, Madras.
- Banerjee, M.N., 1932: A Note on Iron in the Rg Vedic Age, *IHQ*, 8: 364-6.
- Banerjee, N.R., and Soundararajan, K.V., 1950: Sanur - A Megalithic Site in District Chingleput, *Ancient India*, Bulletin of the Archaeological Survey of India (hereafter *A.I.*), Number15, ASI.
- Banerjee, N.R., 1965: *The Iron Age in India*, Delhi.
- Banerjee, N.R., 1966:Amirthamangalam1955: A Megalithic Urn Burial Site in District Chingleput, *A.I.*, 224-36.
- Begbie, L.F., 1908:A *Monograph on the Iron and Steel Industry on the Central Province*, Printed at the Secretariat Press.
- Belliappa, P.K., 1964:Prospects of Iron and Steel Industry in Tamilnadu, Madras Government Press, Madras.
- Benza, P.M., 1836:Geological Sketch of Neilgherries, *Gazetteer, Nilgiris District*, Vol. 422-424.
- Bharadwaj, H.C., 1973: Aspects of Early Iron Technology in India, in D.P.Agrawal and A.Ghosh (Eds), *Radiocarbon and Indian Archaeology*, Tata Institute of Fundamental Research, Bombay, 391-400.
- Bharadwaj, H.C., 1981-83: Studies in Ancient Indian Technology, *Puratattva*, 13-14. 139-48.
- Bhattacharya, Sabyasachi, 1970: Iron Smelters and the Indigenous Iron and Steel Industry of India: From Stagnation to Atrophy, *Journal of Anthropological Society*, 5,137-151.
- Biswas, A.K., 1994: *Minerals and Metals in Ancient India*, D.K.Print World (P) Ltd, New Delhi.
- Bose, D.M., & Subbarayappa, B.V., (eds.), 1971: *A Concise History of Science in India*, Indian National Science Academy, New Delhi.
- Brajdeo, Prasad Roy, The Vedic Spade and Plough, L.N.Mishra Commemoration Volume, *Journal of Bihar Research Society*, 580 -583.
- Britton, S.V., 1934:Indian Iron, *Nature.*, V.134.238-40
- Bronson, B., 1986:The Making and Selling of wootz - A Crucible Steel of India, *Archaeo – Materials*, ICD: 13-51.
- Brooke, J.C., 1864: The Mines of Khatree in Rajpootana, *Journal of Asiatic Society of Bombay*, Vol.33: 519-529. (Hereafter *JOASOB*)
- Brown, J., Coggin and Dey, 1923: *Indian Mineral Wealth*, Oxford.

Bibliography

- Buchanan, F., 1988: *A Journey from Madras*, 3 Volumes, Reprint, Asian Educational Services, Madras/Delhi
- Caley, E.R., 1964: *Analysis of Ancient Metals*, Pergaman Press.
- Campbell, Captain, J., 1842:*Report* Upon the Manufacture of Steel in Southern India, *Report of the Royal Asiatic Society of Bengal*, 217 -218.
- Chakrabarti, D.K., 1974: Beginning of Iron in India: Problem Reconsidered, In A.K.Ghosh (Ed), *Perspectives in Palaeoanthropology - D.Sen Festschrift*, 345- 56
- Chakrabarti, D.K., 1976: The Beginning of Iron in India, *Antiquity*, 50.114-24.
- Chakrabarti, D.K., 1977: Research on Early Indian Iron, 1795-1950,*The Indian Historical Review*, 4.96-105.
- Chakrabarti, D.K., 1979:Iron in Early Indian Literature, *J.A.R.S*, 22-30.
- Chakrabarti, D.K., 1981-83: Study of Iron Age in India, *Puratattva*, 13- 14, 81-85.
- Chakrabarti, D.K., 1985, The Issues of Indian Iron Age, In: S.B.Deo and Paddyya (Eds), *Recent Advances in Indian Archaeology*, 74-77, Deccan College, Pune.
- Chakrabarti, D.K., 1992: *The Early Use of Iron in India*, Oxford University Press, Bombay.
- Chakrabarti, Dilip, K., *1973*: Beginning of Iron and Social Change in India, *Indian Studies, Past and Present*, 14: 329-38.
- Charles, J.A., 1979:From Copper to Iron - the Origin of Metallic Materials, *Journal of Metals*, 30.8-13.
- Chattopadhyay, P.K., 1984: Archaeometallurgical Studies in Indian Sub-Continent: A Survey of Metallography of Iron Objects, *Indian Journal of History of Science*, 9:361-5.
- Chaudhuri, Konark, K., 1988: Ayas in Vedic Literature, In: *Studies in Ancient Indian History*, D.C.Sircar Commemoration Volume, 321-326.
- Cleere, H.F., 1963: Primitive Indian Iron Making Furnaces, *The British Steel Maker*, 154- 8.
- Coghcan, H.H., 1956: *Notes on Pre-History and Early Iron in the World*, Oxford University Press.
- Crackoft, W., 1833: Iron Smelting in the Khasi Hills, *Journal of Asiatic Society of Bengal*, Vol.II.
- Cyril, Stanley, Smith, 1965: *A History of Metallography*, The University of Chicago Press.
- Deo, S.B., 1982: *Excavations at Naikund*, Nagpur, University Press, Nagpur.
- Deo, S.B., 1985: The Megaliths: Their Culture, Ecology, Economy and Technology, *Recent Advances in Indian Archaeology*, 89-99, Deccan College, Pune.
- Deo, S.B., & Paddyya, K., 1985: The Issues of the Indian Iron Age, *Recent Advances in Indian Archaeology:* 74- 77, Deccan College, Pune.
- Deo, S.B., 1994: New Discoveries of Iron Age in India, *Indian Archaeological Heritage*, Agam Kala Prakashan, 189-19, Delhi.
- Dikshit, K.N., 1992: Iron Age and Peninsular India, *Puratattva*, 22, 31-34.
- Elvin, Verier, and Humphry, Milford, 1942: *The Agaria*, Oxford University Press.
- Forbes, R.J., 1950: *Metallurgy* in *Antiquity*, Leiden.
- Forbes, R.J., 1971, *Studies in Ancient Technology*, Leiden.
- Friend, J.N., 1926: *Iron in Antiquity*, London.
- Gaur, R.C., 1983: *Excavations at Atranjikhera*, Motilal Banarasidas, Delhi.
- Ghosh, A., 1974: Iron and Urbanisation in the Ganga Basin, *Indian Historical Review*, Vol. I. 98-103.
- Ghosh, M.K., 1963:The Delhi Iron Pillar and its Iron, *N.M.L. Technical Journal*, Vol.I: 31-45.
- Gogte, V.D., 1982: Discovery of Megalithic Iron Smelting Site by Three - Probe Resistively Survey, *Bulletin of the Deccan College Research Institute*, Vol.40.211-215.
- Gogte, V.D., 1983,Megalithic Iron Smelting at Naikund (part II) Efficiency of Iron Smelting by Chemical Analysis, In: Deo, S.B., & Jamkhedar, A.P., *Naikund Excavations*, Appendix III, 1-4, Deccan College, Pune.
- Gopal, Lallanji, 1960: Antiquity of Iron India, *Uttatra Bharati*, Journal of Andhra Historical Research Series, 28, Vol.9. 71-86.

- Gordon, D.H., 1952: Early Use of Metals in India and Pakistan, *Journal of Royal Anthropological Institute*.
- Gurney, O.R., 1961: *The Hittites*, Harmonworth.
- Gururaja Rao, B.K., 1972:*The Megalithic Culture in South India*, Dharwar.
- Heath, J.M., 1832: Memoranda on Salem Iron Works, *Report of the Royal Asiatic Society of Bengal*, Vol.I. 253-255.
- Heath, J.M., 1837: On Indian Iron and Steel, *Madras Journal of Science and Literature*, Vol.II.185-192.
- Hegde, K.T.M., 1973: A Model for Understanding Ancient Indian Iron Metallurgy, *Man*, Journal of Royal Anthropological Institute, Vol.8 No.3. 416-442.
- Hegde, K.T.M., 1973: Early Stages of Metallurgy in India, *Radiocarbon and Indian Archaeology*, Agrawal, D.P and Ghosh, A., (Eds), Tata Institute of Fundamental Research, Bombay, 401-405.
- Hegde, K.T.M., 1981: Scientific Basis and Technology of Ancient Indian Copper and Iron Metallurgy, *Indian Journal of History of Science*, 16:189- 201.
- Hegde, K.T.M., 1985: Scientific Studies in Indian Archaeology, *Recent Advances in Indian Archaeology*, 101-109, Pune.
- Heyene, B., 1814: *Tracts, Historical and Statistical, on India*, London.
- Holland, T., 1904: *Preliminary Report on the Iron Ores and Iron Industries of the Salem District*, Records of Geological Survey of India, 135-159.
- Hutchinson, A.R., 1950: Iron and Steel Through the Ages, *The Museums Journal*, Vol. 50.
- Joshi, S.D., 1970: *History of Metal Founding on The Indian Sub Continent Since Ancient Times*, Ranchi.
- Kosambi, D.D., 1963: The Beginning of Iron Age In India, *Journal of Economic and Scoial History of Oriental*, 6,Vol.VI.309-318.
- Krishnan, M.S., 1951: *Mineral Resources of Madras*, Calcutta.
- Krishnan, M.S., 1954: *Iron Ore, Iron and Steel*, Calcutta.
- Krishnaswami, V.D., 1949: Megalithic Types of South India, A.I., No.5
- Kuppuram, G., 1989:*Ancient Indian Mining, Metallurgy and Metal Industries*, Vol.II. Sandeep Prakashan, Delhi.
- Lahiri, Dipankar, 1968: Mineralogy in Ancient India, *I.J.H.S.*, Vol.3. 1-8.
- Lal, B.B., 1955: Excavations at Hastinapura, *A.I.*, Nos.10, 11 &12.
- Lamberg, Karlovasky, C.C., 1967: Archaeology and Metallurgical Technology in Pre - Historic Afghanistan, India and Pakistan, *American Anthropologist*, Vol.69, No.2. 145-162.
- Lowe, T.C., 1990: Refractories in High-Carbon Iron Processing: A Preliminary Study of the wootz Making Crucibles, In: Kingery, W.D., (Ed.), Ceramics and Civilisation, Cross-Craft and Cross Cultural Interactions in Ceramics, *The American Ceramic Society*, Vol.4, 237-250, Pittsburgh.
- Maddin, R., 1992:A History of Martensite: Some Thoughts on the Early Hardening of Iron, Olson, In: G.B & Owen.W.S. (Eds), Martensite, *The Materials Information Society*, U.S.A.
- Mahalingam, T.V., 1970: Report *on the Excavations in The Lower Kaveri Valley, Tirukkampuliyur and Alagarai*, University of Madras.
- Mushet, D., 1805: Experiments on wootz, *Philological Transactions*, 163-175.
- Nagaswami.R., 1994: Alagankulam: An Indo-Roman Trading Port, *Indian Archaeological Heritage*, Vol.I. Agam Kala Prakashan, Delhi.
- Narasimhaiah, B., 1980: *Neolithic and Megalithic Cultures in Tamilnadu*, Sandeep Prakashan, Delhi.
- Neogi, P.L., 1914: *Iron In Ancient India*, The Indian Association for the Cultivation of Science, Bulletin, No 12,Calcutta.
- Ojha, A.P., 1981:Black Smith in Ancient and Early Medieval India, *I.H.C.*, Vol.I: 159-162.

Bibliography

- Pearson, G., 1795: Experiments and Observations to Investigate the Nature of a Kind of Steel Manufactured at Bombay and their Called wootz, *Philological Transactions*, 85:322-346.
- Pleiner, R., 1971:The Problem of Beginning of Iron Age In India, *Acta Pre Historica ET Archaeologica*, 2:5-36.
- Prakash, B., 1989-90: Archaeo Metallurgical Study of Iron Pillar at Dhar, *Puratattva*, No.20. 118-122.
- Prinsik, J., 1832:Notes on Salem Iron Works, *J.A.S.B.*, Vol.I. 253-255.
- Purushotam Singh, 1994:The Beginning of Early Iron Age In North India, *Indian Archaeological Heritage*, Agam Kala Prakashan, Delhi.
- Rae, Alexander, 1902-3: Pre-Historic Antiquities in Tinnevelly, *Annual Report of A.S.I.*, Madras, 11-40.
- Ragunatha Rao, B. and Sasisekaran, B., 1997:Guttur-An Iron Age Industrial Centre in Dharmapuri District, *I.J.H.S.*, 32. (4), 347-359. New Delhi.
- Rahman, A., and Subbarayappa, B.V., 1966: A Note on the Native Method of Bar Iron Production in South India (Salem Region), I.J.H.S., Vol.I (2).
- Rajan, K., 1994:*Archaeology of Tamilnadu (Kongu Country)*, Book India Publishing House, Delhi.
- Rajan, K., 1997:*Archaeological Gazetteer of Tamilnadu*, Mano Pathippakam, Thanjavur.
- Raman, K.V., 1968: Arts and Crafts in Ancient Tamilnadu as Revealed in Excavations, *Paper Presented in the 1st world Tamil Conference*, Madras.
- Raman, K.V., 1968: Excavations at Poompuhar, *Handbook*, International Conference on Tamil Studies.
- Raman, K.V., 1968: *Excavations in Kanchi*, Tamil University, Tanjavur.
- Raman, K.V., 1988: *Excavations at Uraiyur* (Tiruchirappalli), Madras University Archaeological Series No.8.
- Raman, K.V., Further Evidence of Roman Trade from Coastal Sites in Tamilnadu, In: Vimala Begley and Richard David De Puma, (Eds), *Rome and the Ancient Sea Trade*, The University of Wisconsin Press.
- Rao, K.N.P., 1991: Delhi Iron Pillar, *Metal News*, Vol. 13, No.5
- Richar, T.A., 1939:Primitive Smelting of Iron, *American Journal of Archaeology*, Vol.XLIV.
- Roland, Kiessling, 1989: Non-metallic inclusions in Steel, *The Institute of Metals*, London.
- Roy, T., 1994: The Ushering of Iron Age in Indian Context, *Indian Archaeological Heritage*, 173-180, Agam Kala Prakashan, Delhi.
- Sahi, M.D.N., 1979: Iron at Ahar, In: *Essays in Indian Proto History*, Agrawal, D.P., and Chakrabarti, (Eds), 365-6,Delhi.
- Sahi, M.D.N., 1980: Origin of Iron Metallurgy in India, *Proceedings of the Indian History Congress*, Bombay.
- Sarada Srinivasan, 1994: wootz Crucible Steel: A Newly Discovered Production Site in South India, *Papers from the Institute of Archaeology*, University College, London.
- Sasisekaran, B., 2002: Metallurgy and Metal Industry in Ancient Tamilnadu - An Archaeological Study, *I.J.H.S.*, 37. (1), 17-29, New Delhi.
- Sasisekaran, B., and Ragunatha Rao, B., 1999: Technology of Iron and Steel in Kodumanal, an Industrial Centre in Tamilnadu, *I.J.H.S.*, 34. (4), 263-272,New Delhi.
- Sasisekaran, B., and Ragunatha Rao, B., 2001: Iron in Ancient Tamilnadu, Metallurgy in India - A Retrospective, NML Golden Jubilee Commemorative Volume, India International Publisher, pp. 92-103, New Delhi.
- Sasisekaran, B., and Ragunatha Rao, B., 2001: Technology of Forge Welding Adopted at Mallappadi - An Iron Age Site in Tamilnadu, *I.J.H.S.*, 36. (4), 91-103,New Delhi.
- Schoff, W.H., 1915: The Eastern Trade of the Roman Empire, *Journal of the American Oriental Society*, New York, Vol.35
- Sharma, A., 1994: Antiquity of Iron at Gufkral, *Recent Advances In Marine Archaeology*, Goa.

- Sharma, G.R., 1960: *The Excavations at Kausambi*, 1957-59, Allahabad.
- Sharma, R.S., 1974: Iron and Urbanisation in the Ganga Basin, *The Indian Historical Review*, I (1): 98-103.
- Singh, S.D., 1962: Iron in Ancient India, *Journal of Economic and Social History of the Orient*, Vol.II, 212-216.
- Soundararajan, K.V., 1994:Kavripattinam Excavations (1963-73), Memoirs of the A.S.I., No 90. New Delhi.
- Subbarayulu, Y., 1984: Vallam Excavations, *Quarterly Research Journal of the Tamil University*, Vol.2.No.4.9-93.
- Subramaniam, B, R., 1964: Appearance and Spread of Iron in India - An Appraisal of Archaeological Data, *J.O.I.*, Vol. 13.No.4.349-359.
- Tylecote, R.F., 1962: *Metallurgy in Archaeology*, Edward Arnold, London.
- Tylecote, R.F., 1967: *History of Metallurgy*, Edward Arnold, London.
- Tylecote, R.F., 1968: *The Solid Phase Welding of Metals*, Edward Arnold, London.
- Udayaravi, Moorti, 1994: *Megalithic Culture of South India – Socio-Economic Perspective*, Ganga Kaveri Publishing House, Varanasi.
- Vaithialingam, S., 1977: *Fine Arts and Crafts In Pattu-p-pattu and Ettu-t-tokai*, Annamalai University.
- Varma, K.C., 1982:The Iron Age, the Veda and the Historical Urbanisation, *Indian Archaeology-New Perspective*.
- Vibha, Tripathi, From Copper to Iron, *Puratattva*, 15.75-79.
- Vibha, Tripathi, 1973: Introduction of Iron in India, *Radiocarbon and Indian Archaeology*, Tata Institute of Fundamental Research, Bombay, 272. -4.
- Vibha, Tripathi, 1981: Antiquity of Iron in Madhya Pradesh, *Indian Archaeology - New Perspectives:* 257-262.
- Vijay Kumar Thakur, 1980: Differences Between Bronze Age and Iron Age Cities of Ancient India, *Proceedings of the Indian History Congress*, Bombay.
- Voysey, H., 1832: Description of Native Manufacture of Steel in South India, *Journal of Royal of Royal Asiatic Society, Bombay*, 245-7.
- Wain, Wright, G.A., 1936: The Coming of Iron, *Antiquity*, Vol. X 5-24.
- Warmington, E.H., 1928:*The Commerce Between the Roman Empire and India*, Cambridge.
- Wood, C., 1893:Discussion of Turner's Paper on Production of Iron in Small Furnaces in India, *Journal of Iron and Steel Institute*, 44 (2). 177, 80.
- Wynne, F.H., 1904: Native Methods of Smelting and Manufacture Iron in Jabalpur, Central Province, *J.I.S.I.*, 578.

Select Index

- Adichanallur, Mining Activity at: theory of, by Badrinarayan and Rao, 34
- Banerjee, N.R.: on introduction of iron in South India, 4
- Buchanan, Francis: on furnace used for smelting, 4; on method of making steel, 4; survey of iron industry by, 4;
- Chakrabarthi, Dilip: on independent origin of iron in India, 4
- Early Indian Iron, previous research on: 1-7
- Edax Analysis; of iron bar from Mallappadi, 40, 41
- Furnace Technology: Balfour on, 22, 23; Banerjee on, 25; Buchanan on, 23; Fischer on, 24, 25; Heath on, 24; Hegde on, 21; Heyene, Dr., on, 23; Joshi on, 21; techniques, types, & methods of, 21-25; Tylecote on, 21; Voysey on, 23
- Furnaces, vestiges of, at: Dhatwa, 16; Guttur, 17,18; Jodhpura, 16; Naikund, 16; Kattankulathur, 20; Kodumanal, 18, 19; Mel-Siruvalur, 19, 20; Perungalur, 20; Tiruvarangulam, 20, 21; Veppangudi, 21
- Furnace: bowl type, at Kodumanal, 18
- Indian Steel: export of, to Damascus, Irenopolis, Abyssinian & African ports, 3
- Iron Age in Tamilnadu: date and features of, 11-13
- Iron and Steel: ancient production centers, 16-21
- Iron Industry: Literary reference to, 13-15
- Iron Objects: typology of, from iron age sites: 31-35
- Iron Ore in Tamilnadu: distribution of, 7-11
- Metallurgical Studies of, iron objects from: Guttur, 36-38; Kanchipuram, 54, 57; Kodumanal, 44-54; Mallappadi, 38-44; Mel-Siruvalur, 57
- Mushet, David: on making wootz by fusion technique, 4
- Neogi, Panchanan: analysis and study of iron objects by, 3; on Aryan settlers knowledge of iron, 3
- Pearson: erroneous theory by, on steel making method, 4
- Steel Production Techniques: Bhardwaj on 25; Developments in , 25-29; Foote on, 25; Pacey, Arnold, on 26
- Twin Furnaces: at Guttur, 13; at Malabar region, 18
- wootz Steel: at Godavari, 29; at Tiruvannamalai region, 28; blades of, at Sheffield, 26; Buchanan on furnace used for, 27; by carburisation process, 5; by Deccani or Hyderabadi Process, 5, 26, 28; by fusion process, 4; Chakrabarti, on, 28, 29; Damascus swords from, 25; Heath on, 27; in South Arcot, 5; Mushet, David, on, 4; Mysore Process in making, 5; production of, at Mel-Siruvalur, 20; Production of, at Salem & Tiruchirappali region, 5; Salem process in making, 5; Sarada Srinivasan on making, 5; use of crucible process in, 29-30; use of flux in making of 29; use of rice husk in making, 27
- wootz: derived from Tamil *urukku* or *ukku*, 1

List of Plates

1.	Stone Anvil from Kodumanal.	1:1
2.	General View of the Industrial Site Guttur, Dharmapuri District.	1:2
3.	Exposed Portions of the Twin Elongated Oval Furnace, Guttur.	1:3
4.	Structure Revealing Three Openings with Earthen Pipes, Guttur.	2:1
5.	Side View of the Furnace, Guttur.	2:2
6.	General View of the Iron Age Industrial Site, Kodumanal, Erode District.	3:1
7.	wootz Steel Ingot from Pakkam, SouthArcot District.	3:2
8.	General View of the Industrial Site, Pakkam.	3:3
9.	Vitrified wootz Crucible, Melsiruvallur, SouthArcot District.	3:4
10.	General View of the Industrial Site, Melsiruvallur.	3:5
11.	Exposed Portions of the Iron Furnace, Kattankulathur, Chingleput District.	3:6
12.	Pits Used to Crush Iron Ore Near the Furnace, Kattankulathur.	4:1
13.	Bowl Furnace, Perungalur, Pudukkottai District.	4:3
14.	General View of the Industrial Site, Veppangudi, Pudukkottai District.	4:4
15.	Iron Blade, Mallappadi, Dharmapuri District.	4:2
16.	Iron Sword, Kodumanal.	5:4
17.	Horse Stirrup, Kodumanal.	5:1
18.	Iron Objects, Kodumanal.	5:2
19.	Iron Sickle, Kaveripattinam, Tanjavur District.	5:3
20.	Adichanallur Burial Site, Adichanallur, Tirunelveli District.	6:1
21.	Industrial working, Mined area, Adichanallur Burial Site.	6:2
22.	Iron Sword, Adichanallur.	7:1
23.	Iron Dagger, Adichanallur.	9:1
24.	Iron Trident, Adichanallur.	7:2
25.	Iron Axe, Adichanallur.	8:1
26.	Iron Hoe, Adichanallur.	8:2
27.	Iron Saucer Lamps, Adichanallur.	9:3
28.	Iron Tripod, Adichanallur.	10:1
29.	Iron Slag, Guttur.	7:3
30.	Iron Artefact, Guttur.	7:4
31.	Microstructure Consisting of Cementite (white), Fine Pearlite (Black) and leduburite, 200X.	12:1
32.	Microstructure Showing Leduburite, 500X.	12:2
33.	Microstructure Showing Platelet of Primary Cementite and Secondary Cementite along Grain Boundaries in the Matrix of Fine Pearlite, 1000X.	12:3
34.	Primary Cementite Platelet with the Matrix Being Fine Pearlite, 100X.	13:1
35.	Microstructure Revealing Acicular Martensite, 100X.	13:2
36.	Structure Showing Cracks in the Cementite and Martensite Phases, 50X.	13:3
37.	Mallappadi General View.	11:2
38.	Macro Figure of the Iron Bar, Mallappadi, Longitudinal Cross Section, Side View.	10:2

List of Plates

39.	Widmanstatten Structure, 100X.	14:1
40.	Structure showing Manganese sulphide inclusions, 500X.	14:2
41.	Structure Showing Pearlite Grains, 100X.	15:1
42.	Pearlitic Structure, Higher Magnification, 1000X.	15:2
43.	Microstructure Revealing Brownish Corrosion Product, 100X. Longitudinal Cross Section, Top View.	16:1
44.	Microstructure in the Longitudinal Surface (top) Showing the Interface Between the Two Metals, A&B, 100X.	16:2
45.	Thin Layer of Slag found at the Interface of the Metals B&C, 100X.	17:1
46.	Widmanstatten Structure Containing Mostly Ferrite and Few Pearlite, 100X.	17:2
47.	Equiaxed Grains of Ferrite, Region C, 100X.	18:1
48.	Cementite and Iron Carbide Distributed as Thin Needles, 100X.	18:2
49.	Equiaxed Grains of Ferrite with Pearlite, 200X. Transverse Cross Section	19:1
50.	Structure with a Demarcation line Separating the Three Metals, 100X.	19:2
51.	Iron with Slag Inclusions, 200X.	20:1
52.	Structure Showing Ferritic grains, 200X.	20:2
53.	Barbed and Socketed Arrowhead, Kodumanal.	7:6
54.	Ferritic Grains with Few Small Grains of Pearlitic Structure, 100X.	21:1
55.	Ferritic Grains with Pearlitic Structure at Higher Magnification, 200X.	21:2
56.	The Same Structure at a Higher Magnification of 500X.	22:1
57.	Net work of Ferrite in a Pearlitic Matrix, 200X.	22:2
58.	Lamellar Pearlite, Ferrite and Cementite, 200X.	23:1
59.	Microstructures Containing Darkly Etched Pearlite Surrounded by Lightly Etched Cementite, 200X.	23:2
60.	Corrosion Pits, 200X.	24:1
61.	Slag Entrapments, 50X.	24:2
62.	Leaf Shaped Arrowhead, Kodumanal.	9:4
63.	Darkly Etched Pearlite is seen alongwith Lightly Etched Cementite, 100X.	25:1
64.	Ferritic Grains with a Few Grains of Pearlite, 500X.	25:2
65.	Pearlite is not Resolved, 1000X.	26:1
66.	Elongated Streak of Sulphide Inclusions, 100X.	26:2
67.	Pearlite with a Few Grains of Ferrite, 500X.	27:1
68.	The Microstructure of the Slag Region Showing Silica, 100X.	27:2
69.	Iron Chisel, Kodumanal.	9:5
70.	Structure Containing Mostly Pearlite, 1000X.	28:1
71.	Structure Showing Network of Ferrite, 200X.	28:2
72.	Lamellar Pearlitic Structure Interwoven in a Network of Lightly Etched Cementite, 500X.	29:1
73.	Sword Bit, Kodumanal	8:3
74.	Microstructure of Ductile Iron with Graphite Nodules (Spheroidite) in an Envelope of Free Ferrite, 200X.	29:2
75.	Iron Dagger, Kodumanal.	9:2
76.	Low Carbon Steel Region, 1000X.	30:1
77.	The Cutting Edge Showing Forge Welded Region, 200X.	30:2
78.	High Carbon Region in the Cutting Edge, 1000X.	31:1
79.	Darkly Etched Pearlitic Steel Interwoven with a Network of Lightly Etched Ferrite, 200X.	31:2

80. The Centre Portion of the Dagger Showing Ferritic Grains in the Micro Structure with a Few Patches of Pearlite, 500X. — 32:1
81. Cutting Edge Showing Mostly Pearlitic Grains with Globular Carbides, 1000X. — 32:2
82. Corroded Portion of the Cutting Edge, 100X. — 33:1
83. White Coating Layer in the Cutting Edge, 100X. — 33:2
84. Iron Nail, Kodumanal. — 7:5
85. Microstructure Consisting of Single Phase (Alpha) Grains of Ferrite, 100X. — 34:1
86. Pearlitic Structure, 1000X. — 34:2
87. Darkly Etched Fine Lamellar Pearlitic Structure Surrounded by Lightly Etched Cementite in the Nail Piece, 200X. — 35:1
88. Zones Containing Wrought Iron Revealing Equiaxed Grains of Ferrite and Elongated Streaks of Slags, 200X. — 35:2
89. Regions Containing Oxide and Silicate Slag Layers, 100X. — 36:1
90. Structure of Slag Reveals Solidification of the Slag by Fast Cooling During forging operation, (fabrication of the nail), 100X. — 36:2
91. Iron Bead, Kodumanal. — 11:1
92. The macro structures of the transverse cross-section of the artifact. — 10:3
93. Equiaxed ferritic grains and elongated streaks of slag inclusions, 50X. — 37:1
94. The grain size of the ferrite in the grain coarsened region, 100X. — 37:2
95. Fine re-crystallised grains found nearer to the inner surface of the iron bead, 100X. — 38:1
96. Fine re-crystallised grains found closer to the outer surface of the iron bead, 100X. — 38:2
97. Ends of the plates showing forge welded joint, 50X. — 39:1
98. The presence of slag and oxide entrapment in the welded joint region, 100X. — 39:2
99. Slag and oxide entrapment in the welded joint region at higher magnification, 500X. — 40:1
100. Iron Tang, Perur. — 4:5
101. X-ray radiograph of Iron hook – Kanchipuram. — 9:6
102. Denteritic structure in slag from Kodumanal, 200X. — 40:2

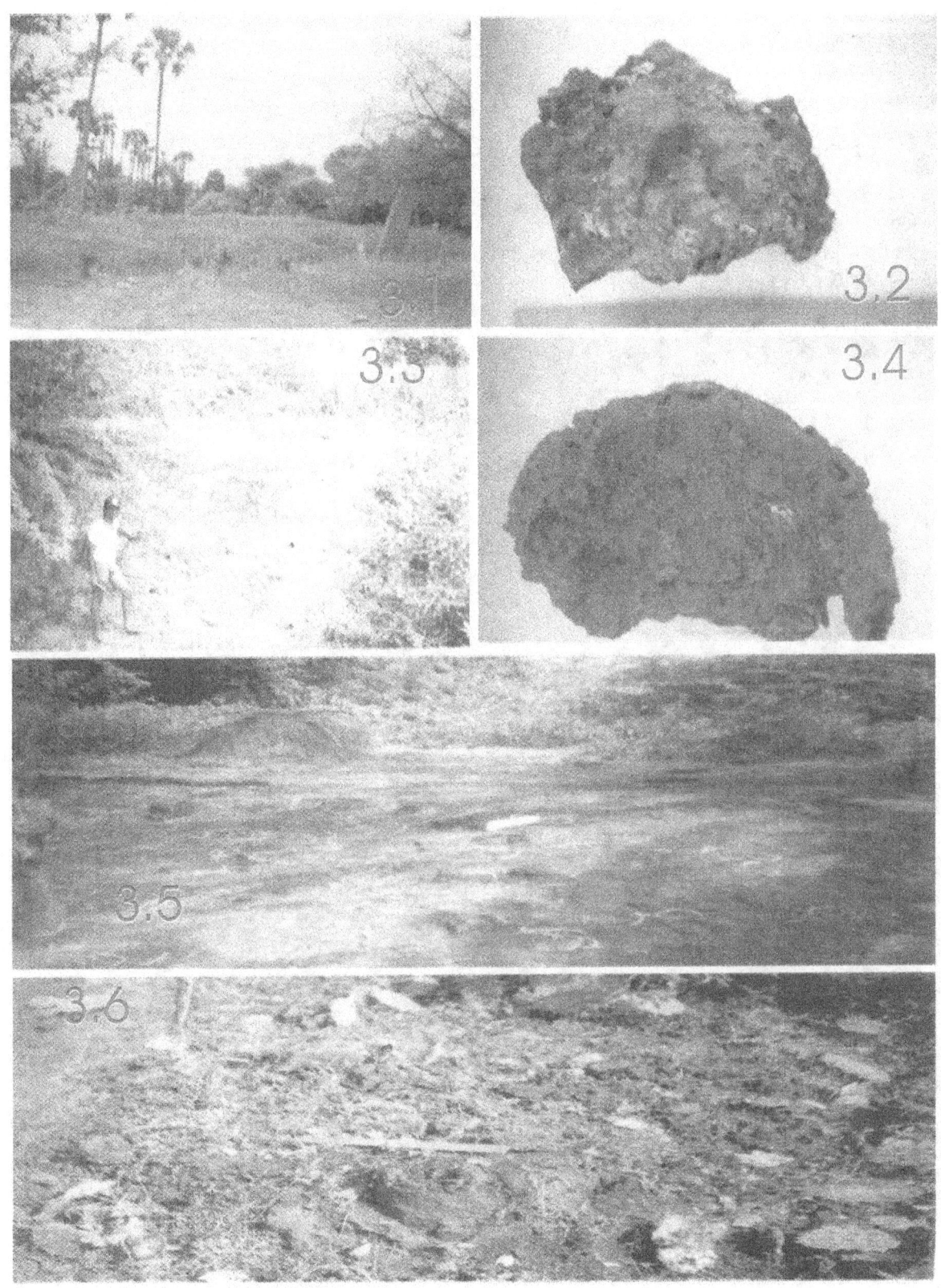

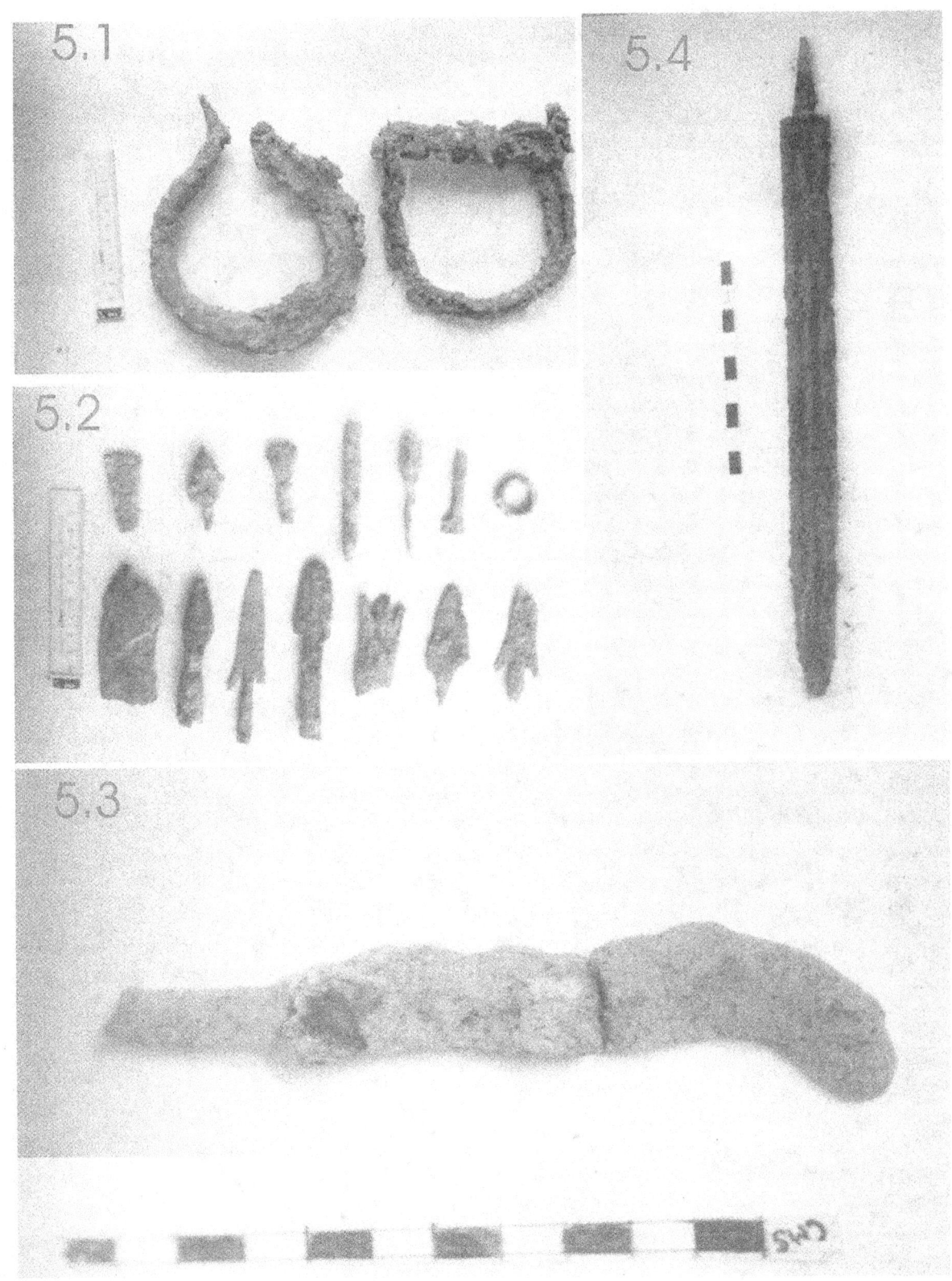

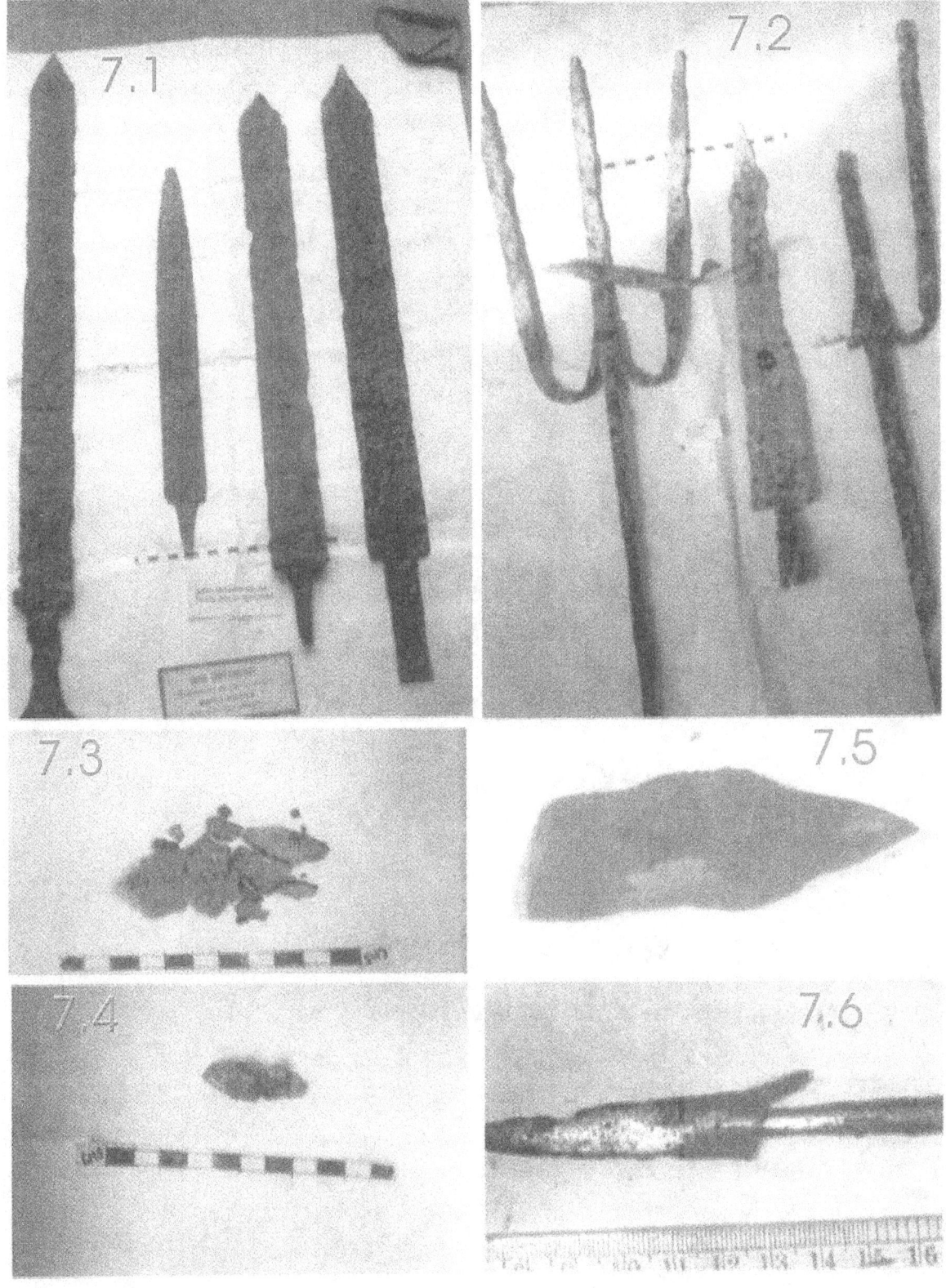

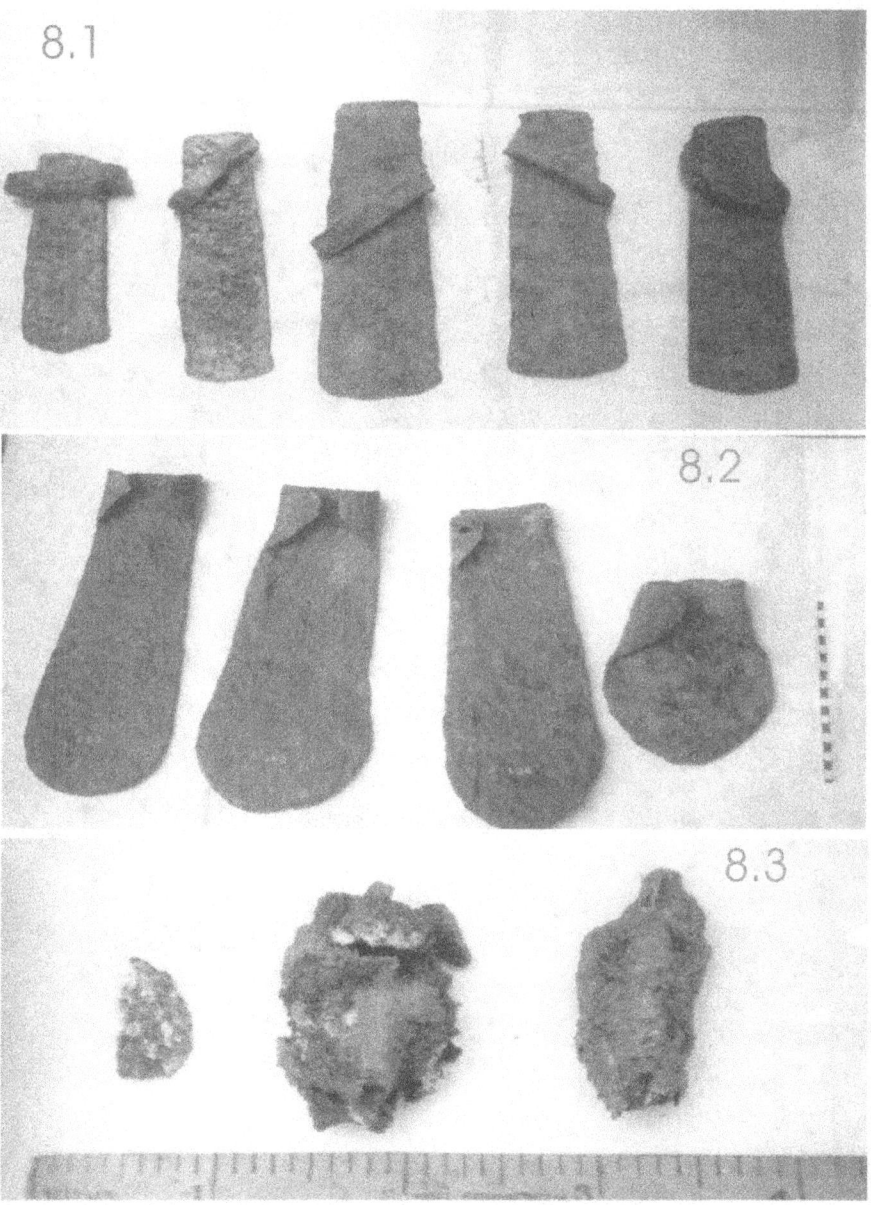

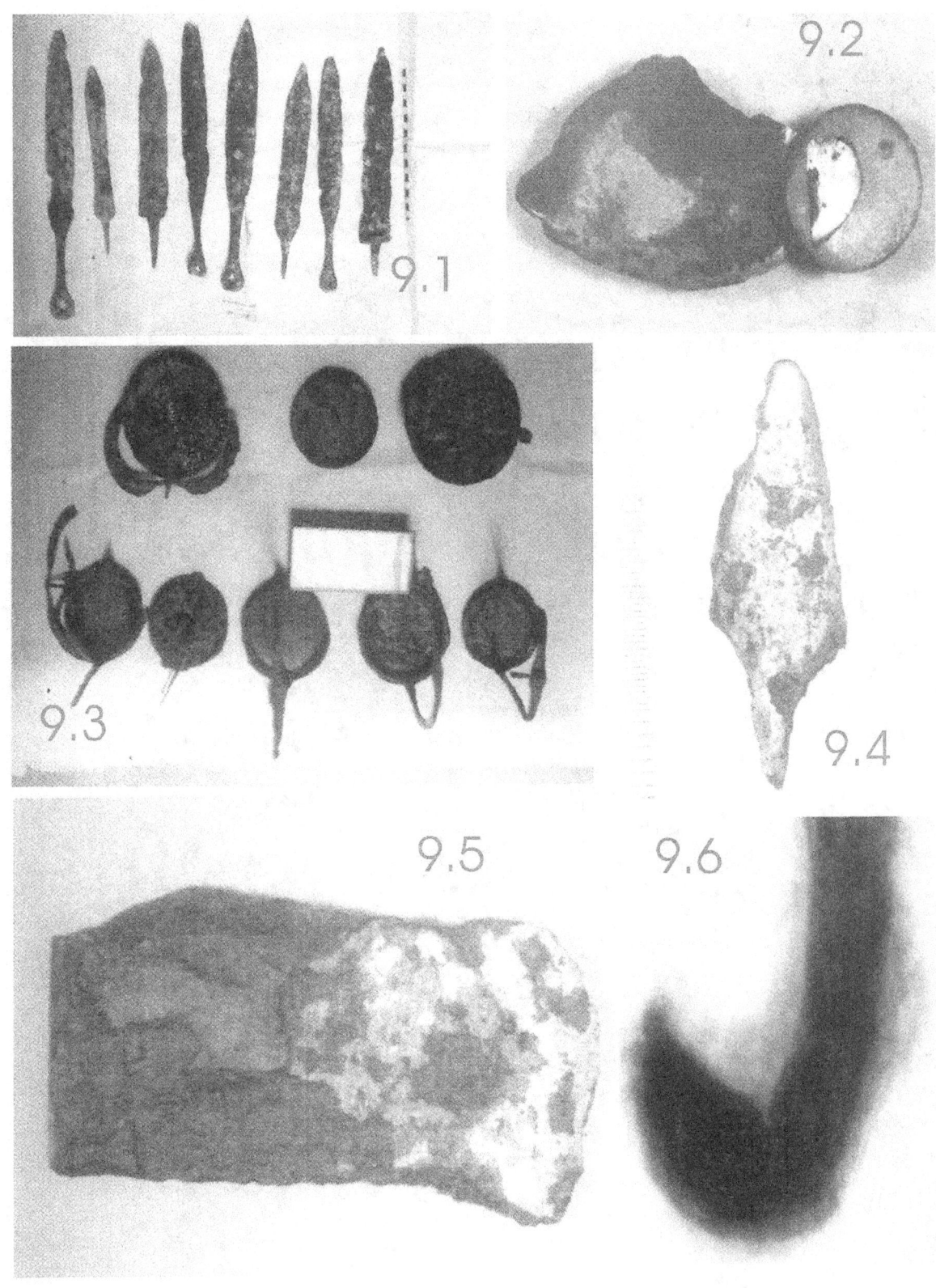

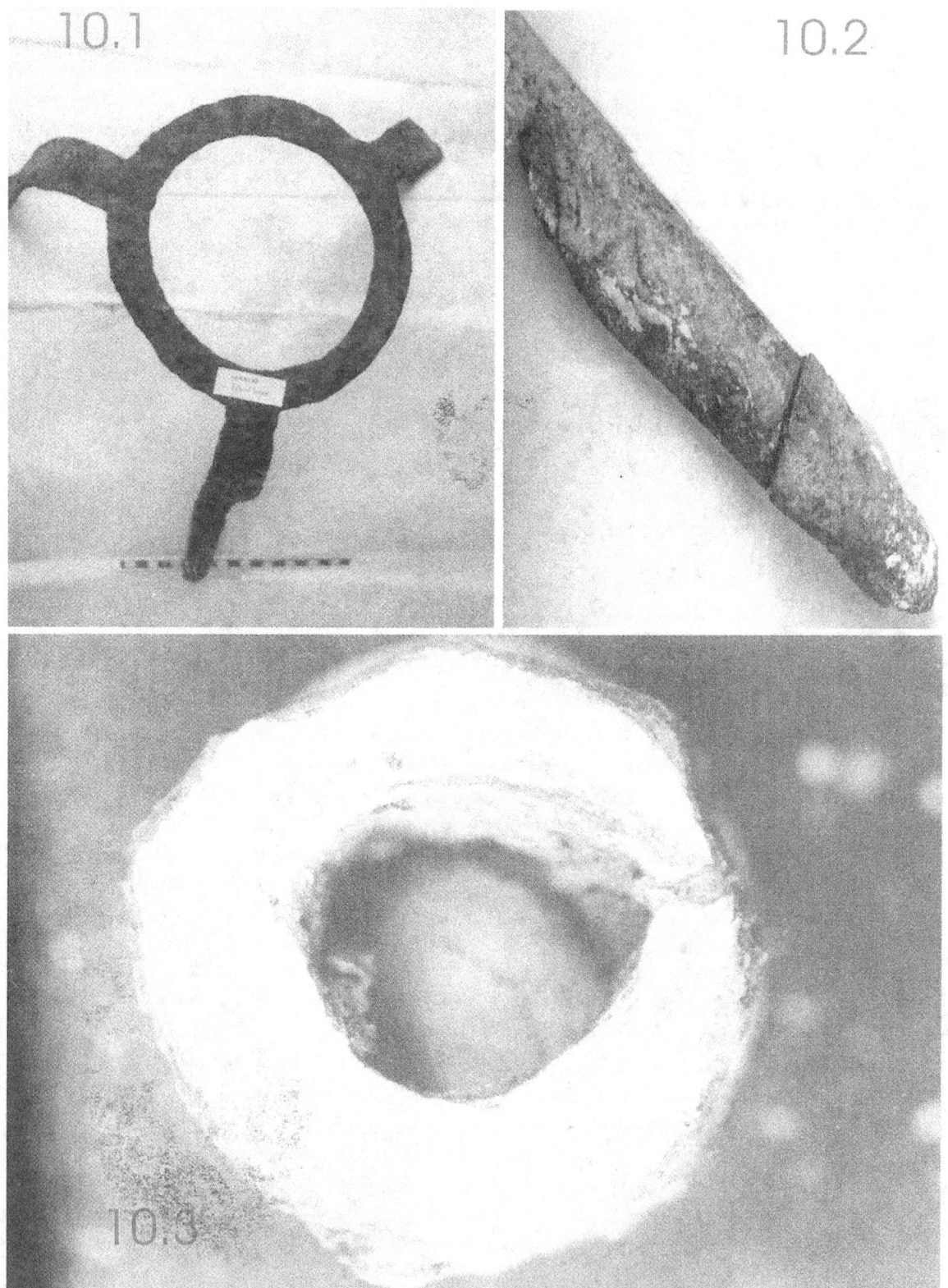

11.1

11.2

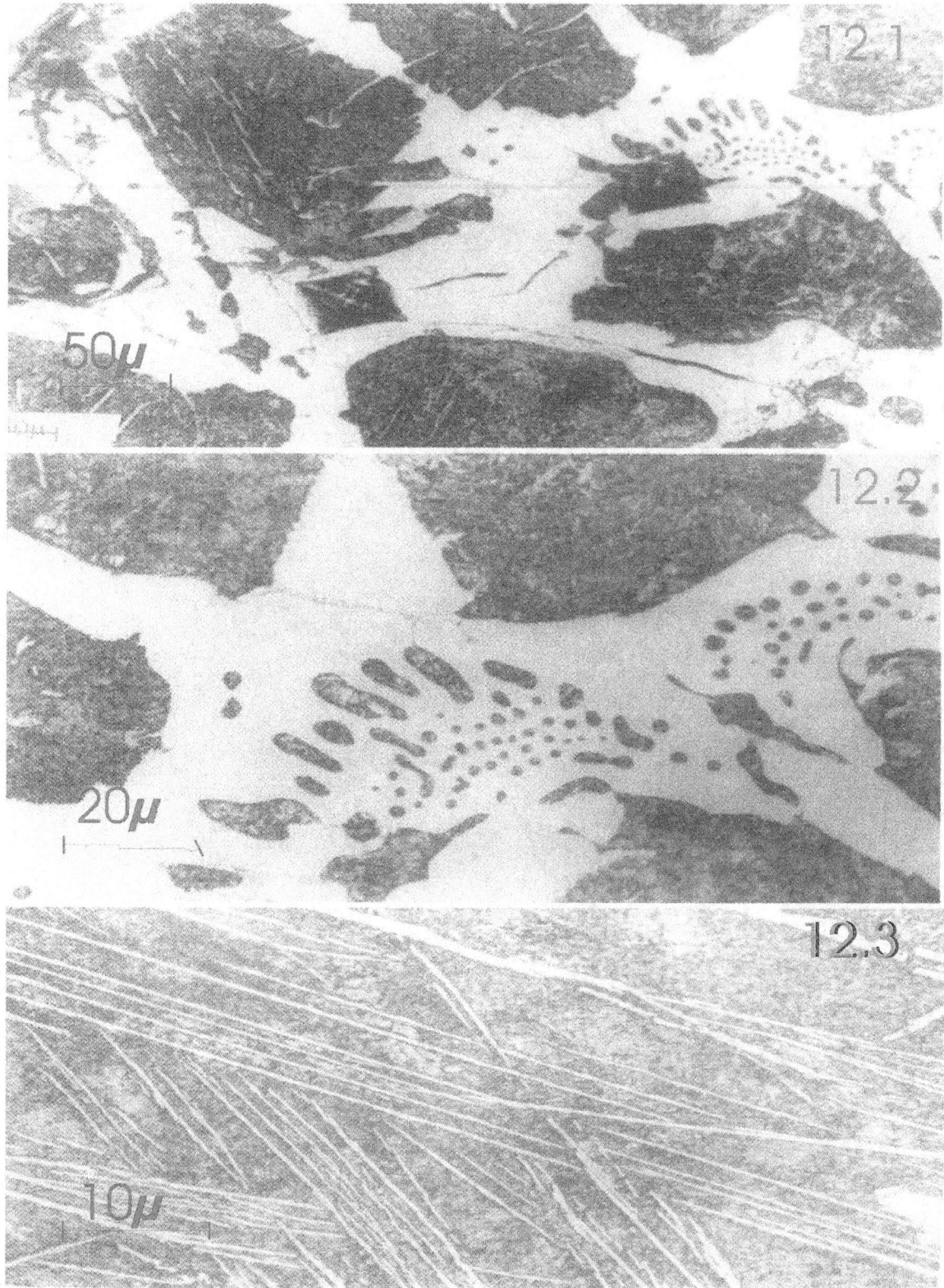

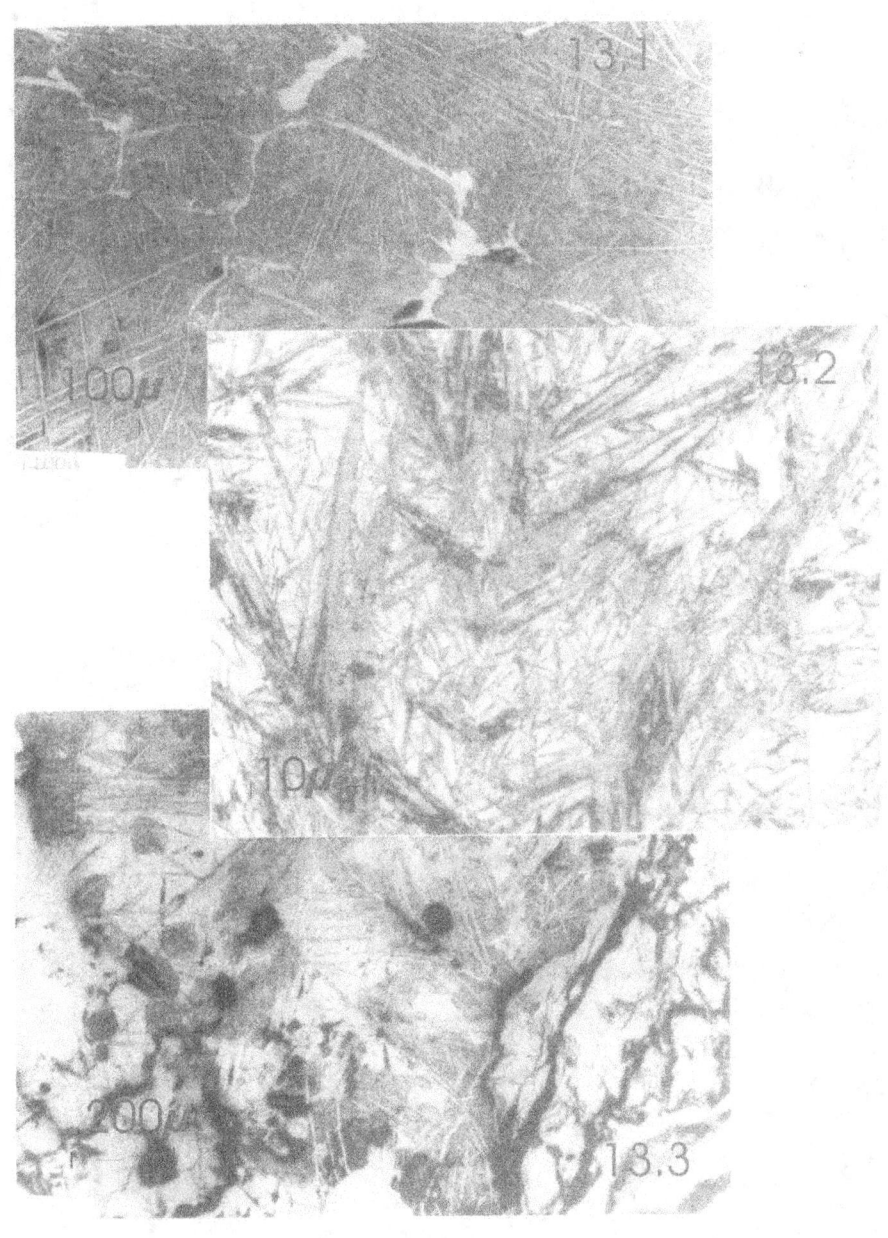

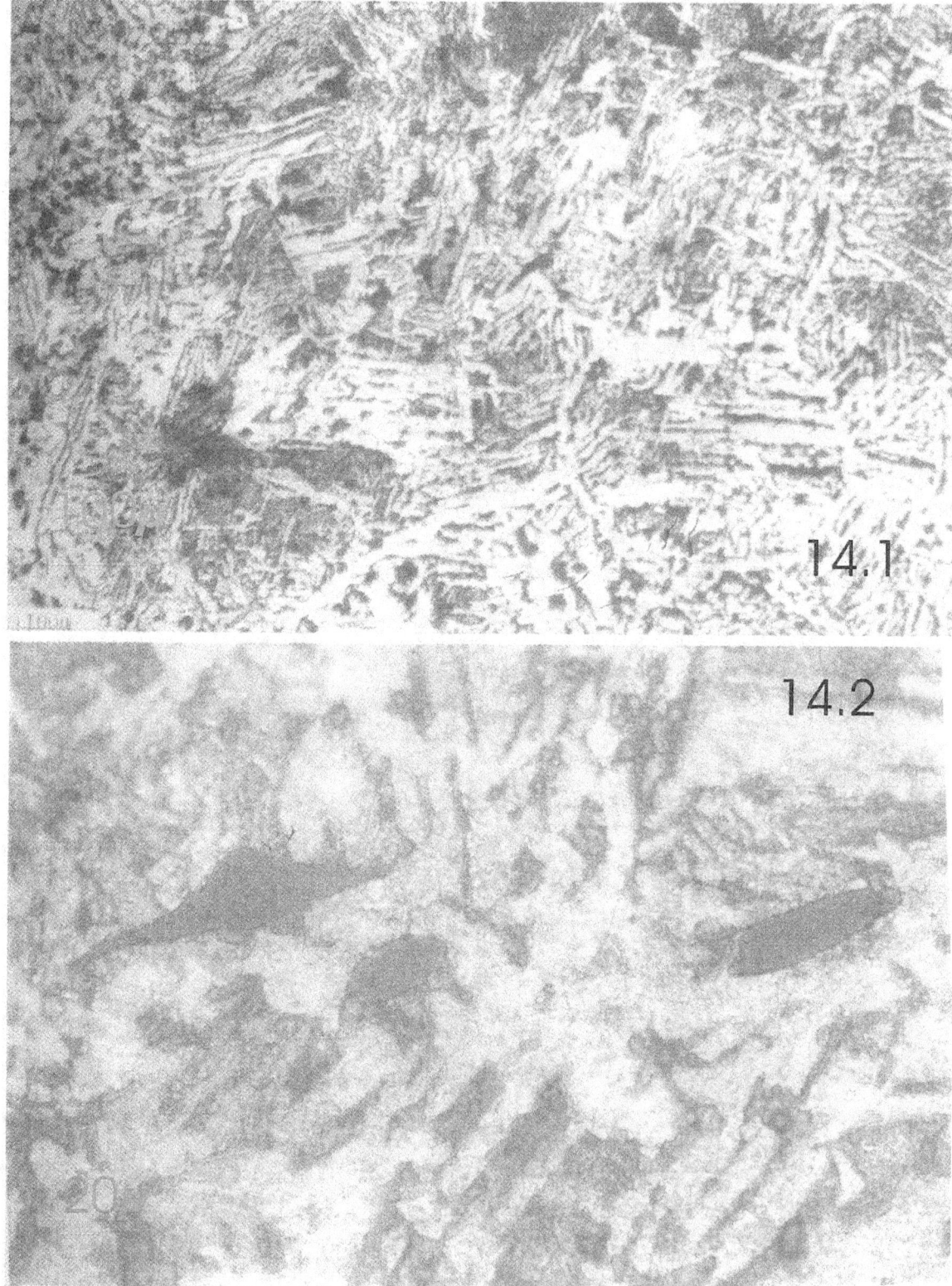

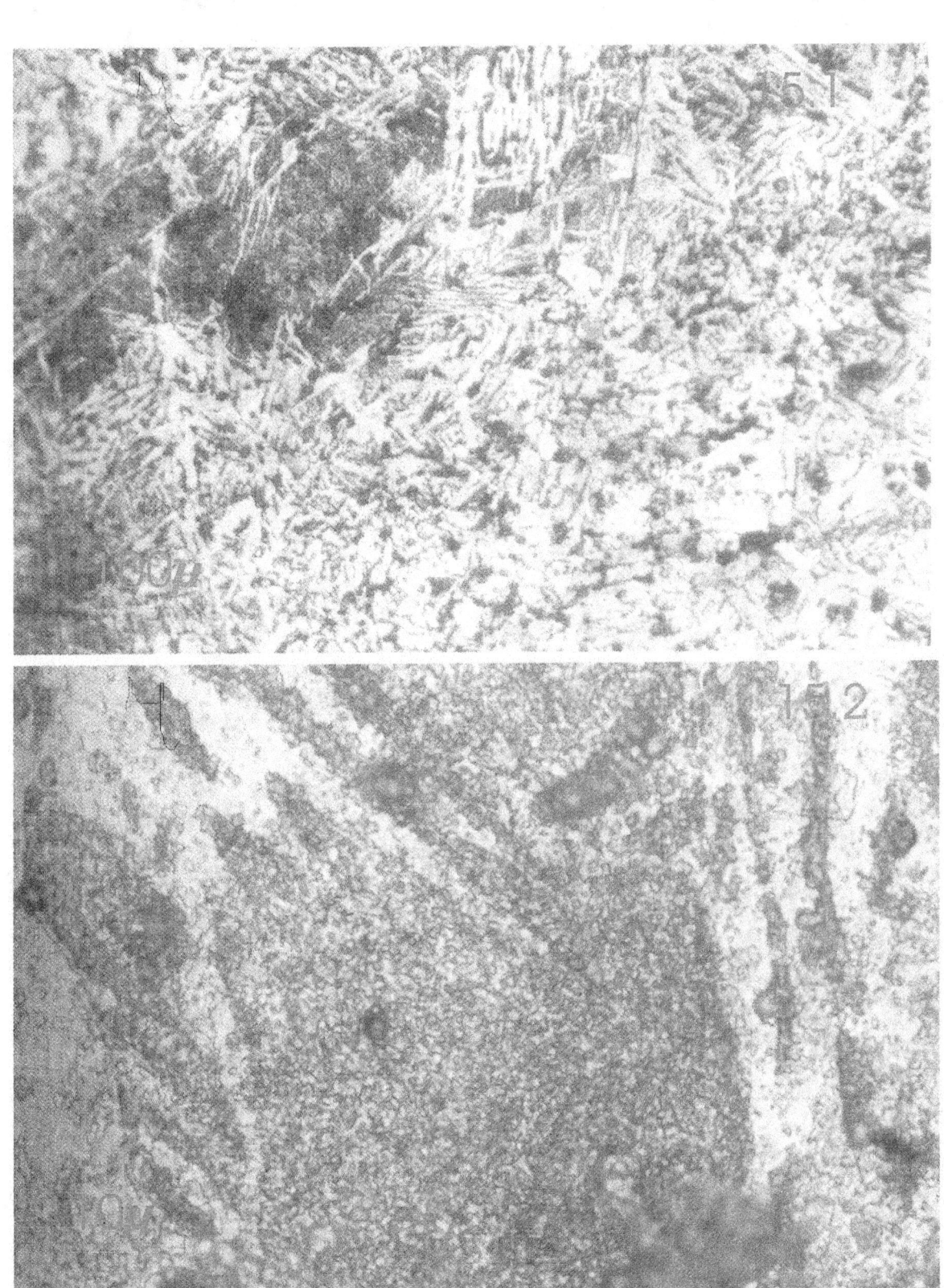

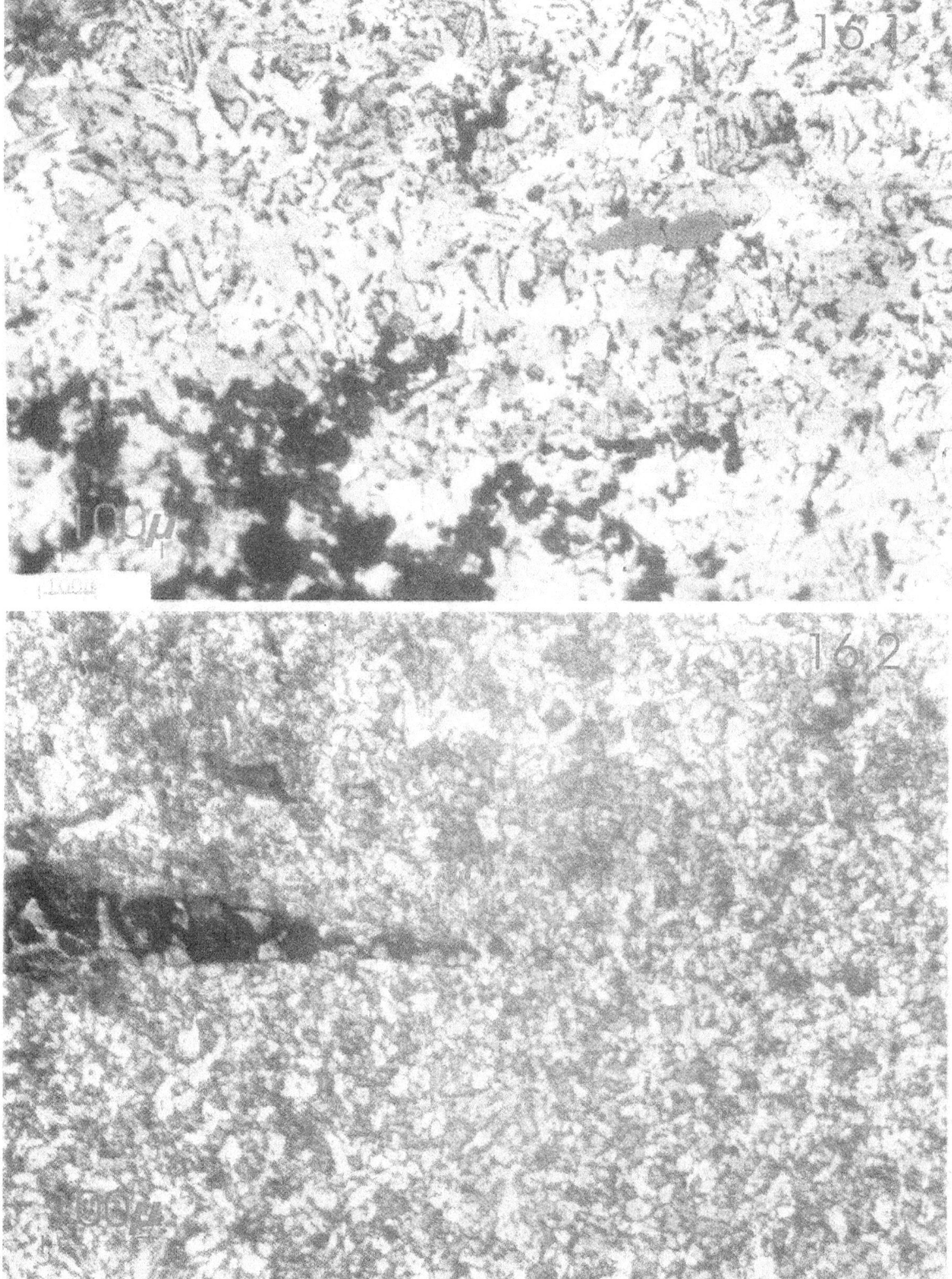

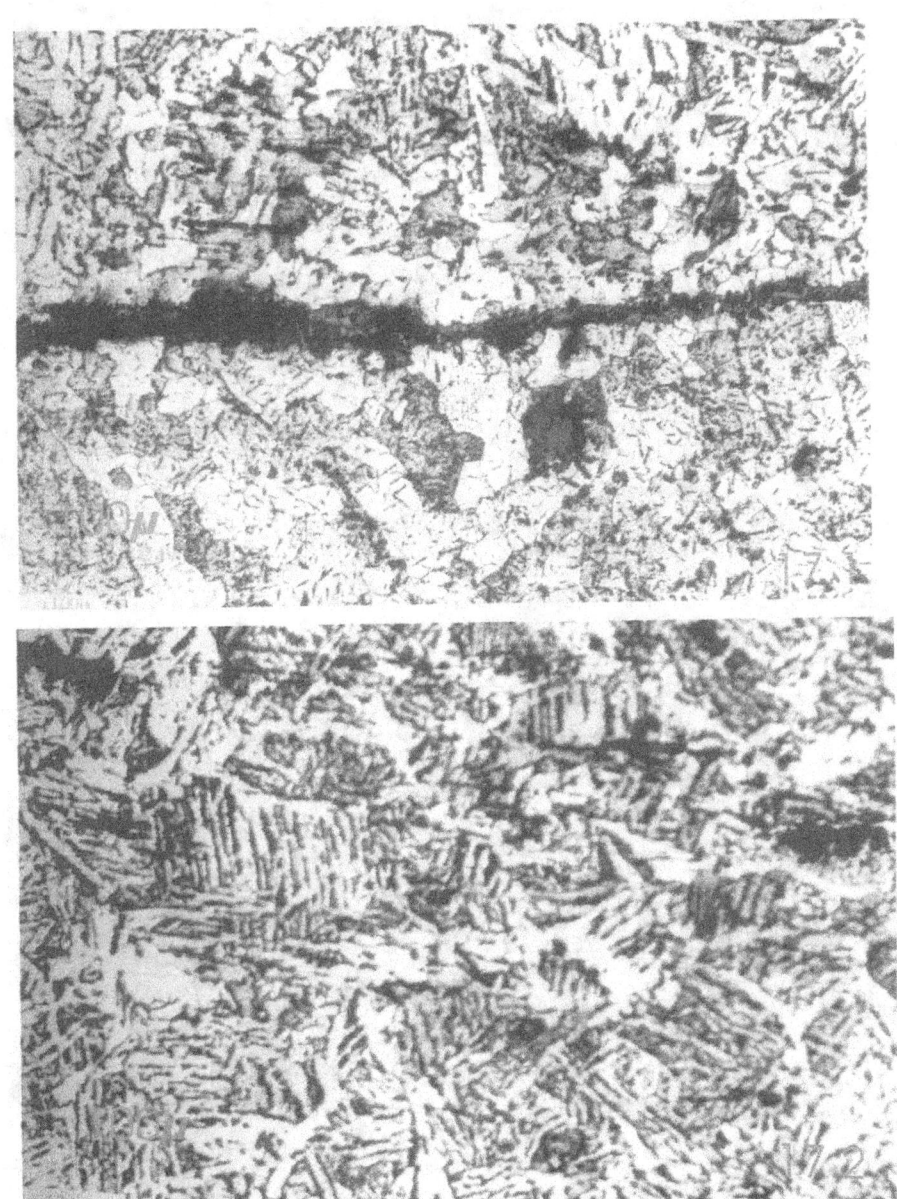

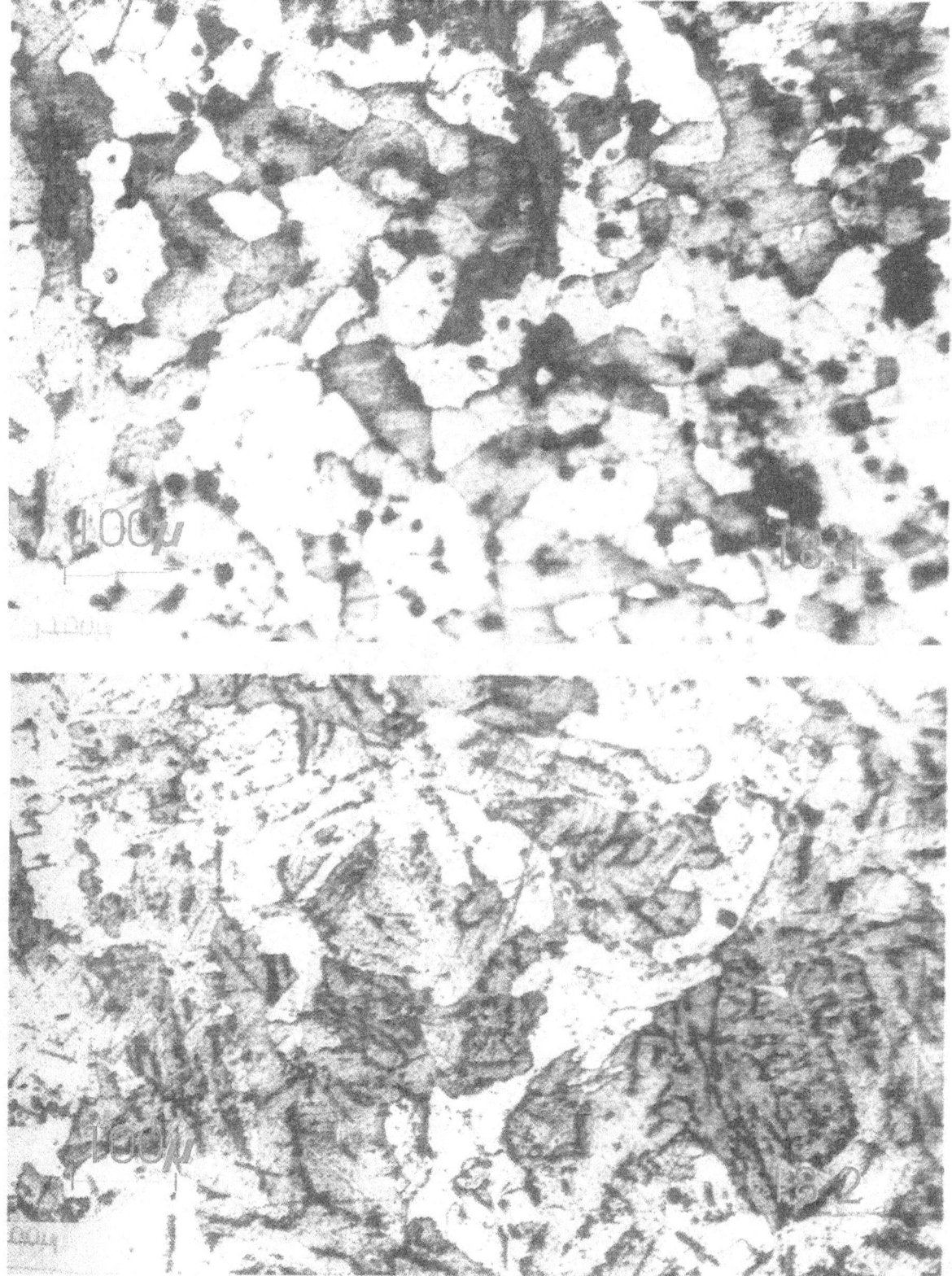

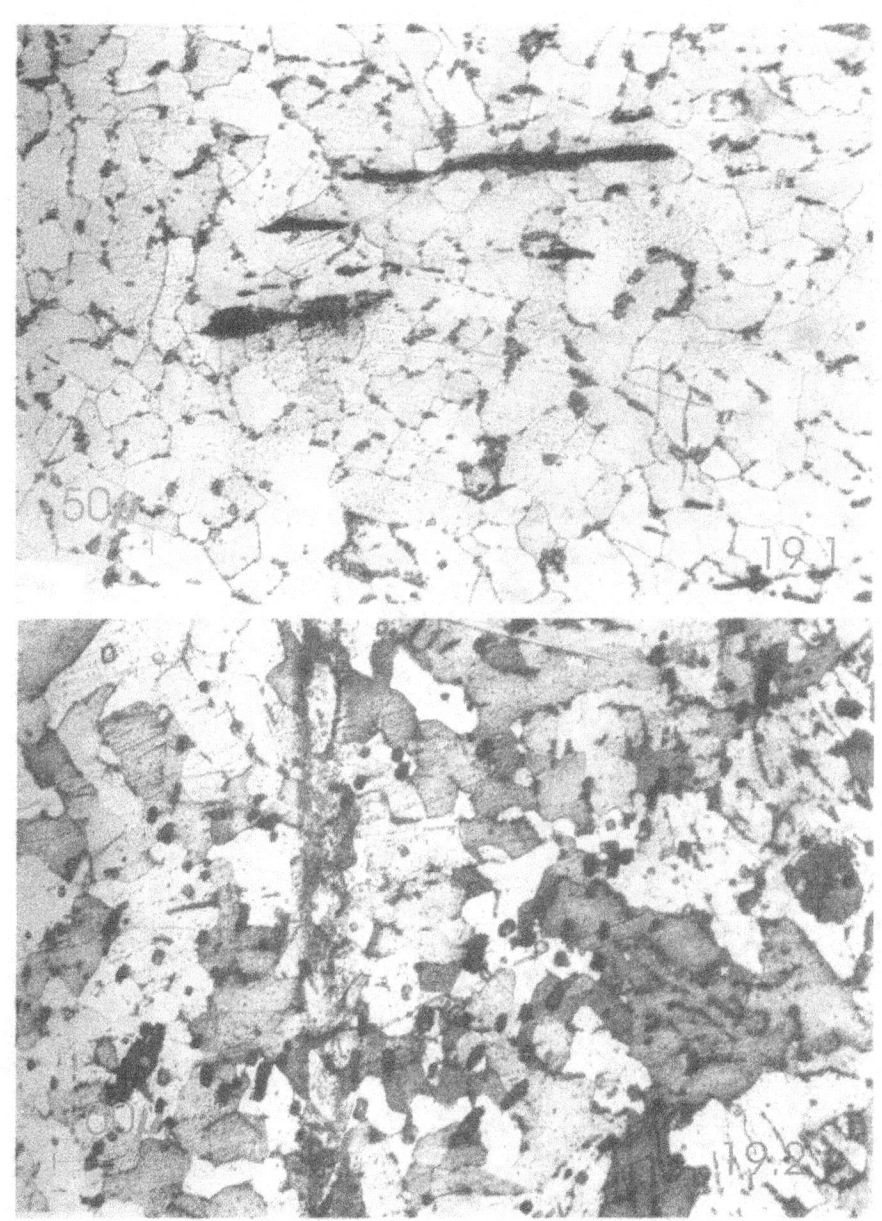

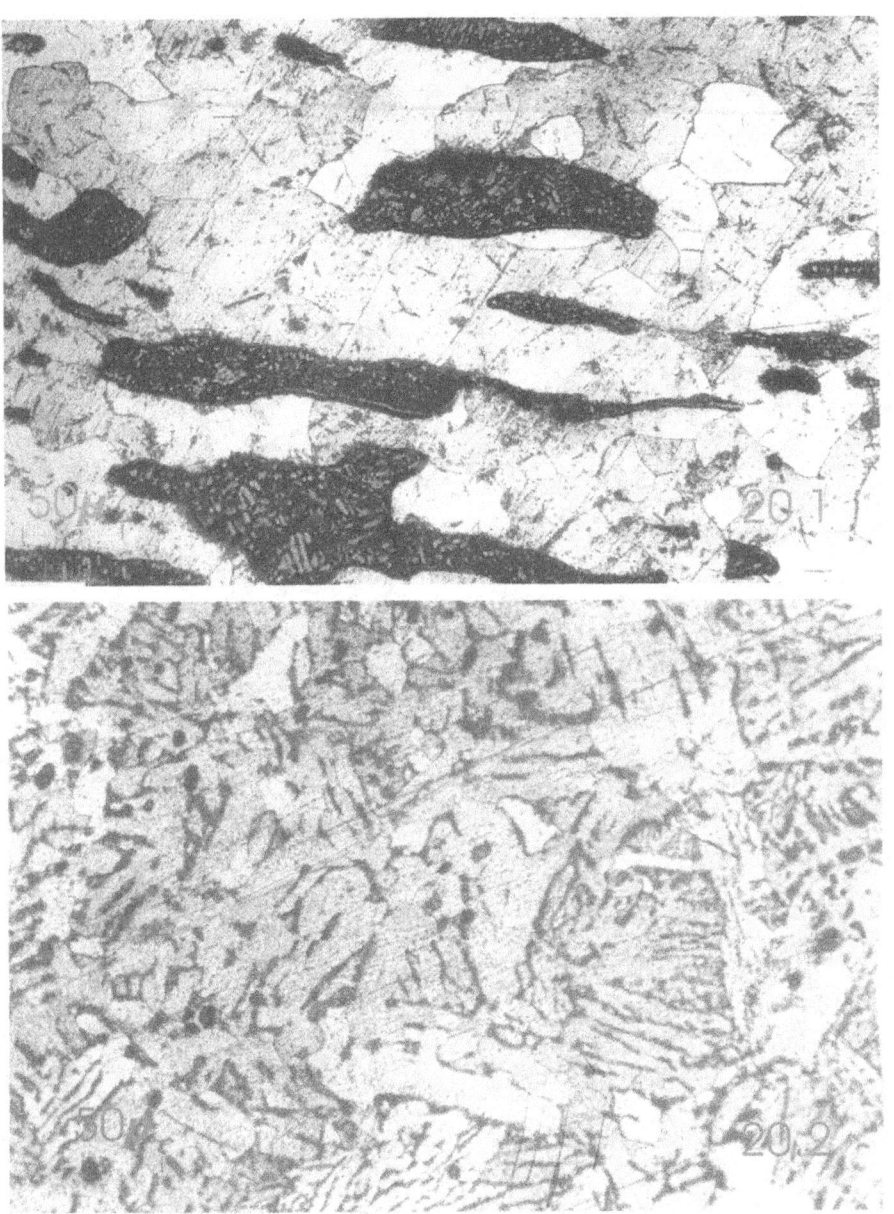

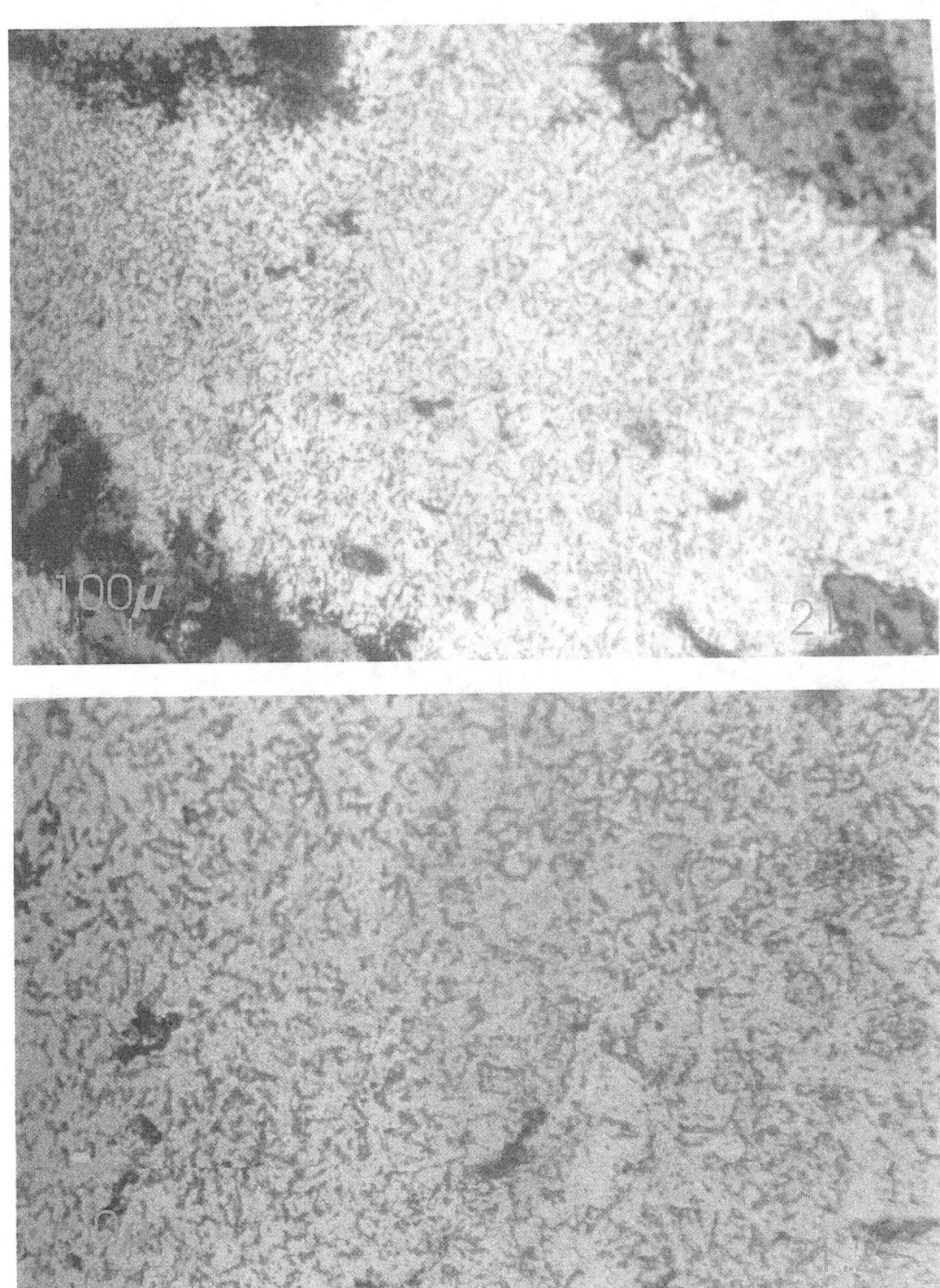

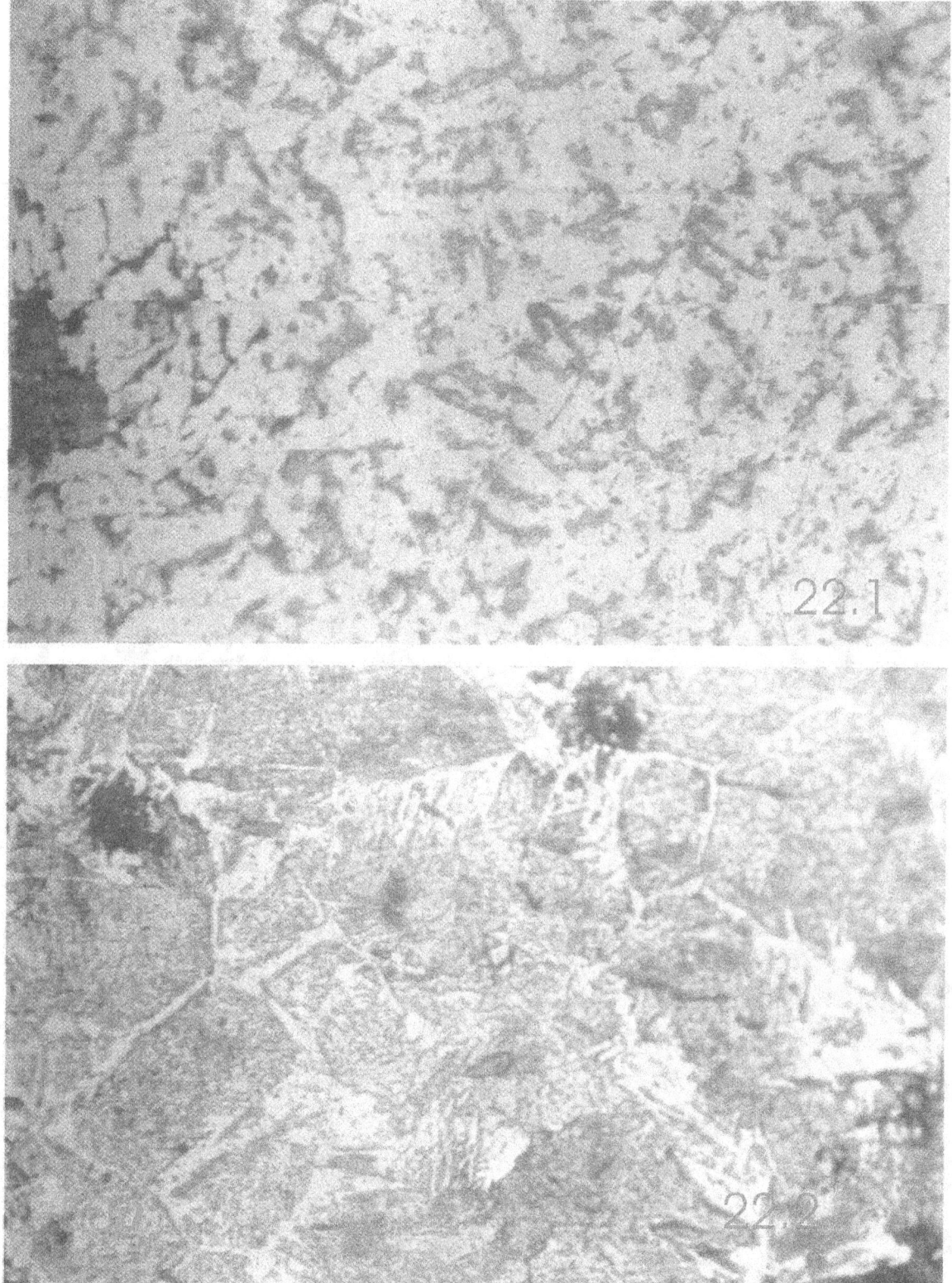

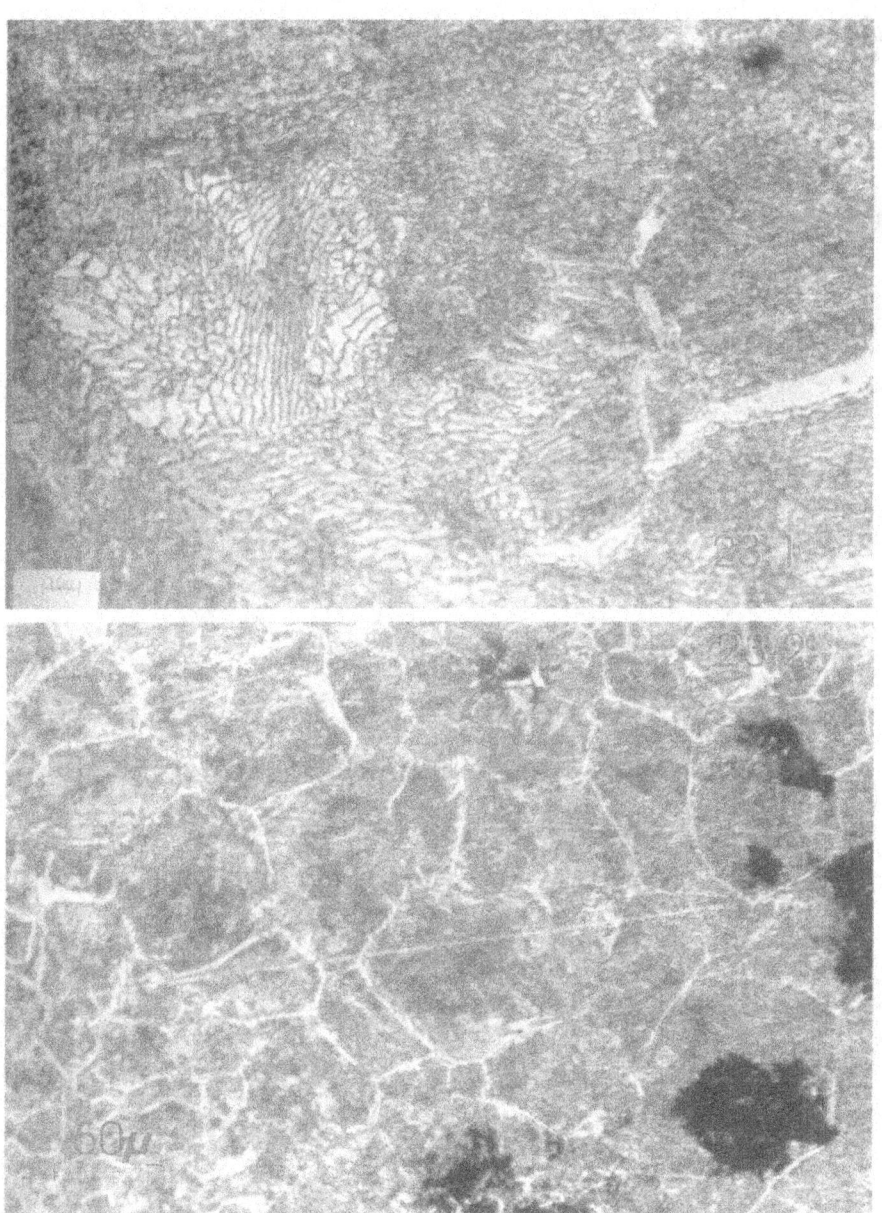

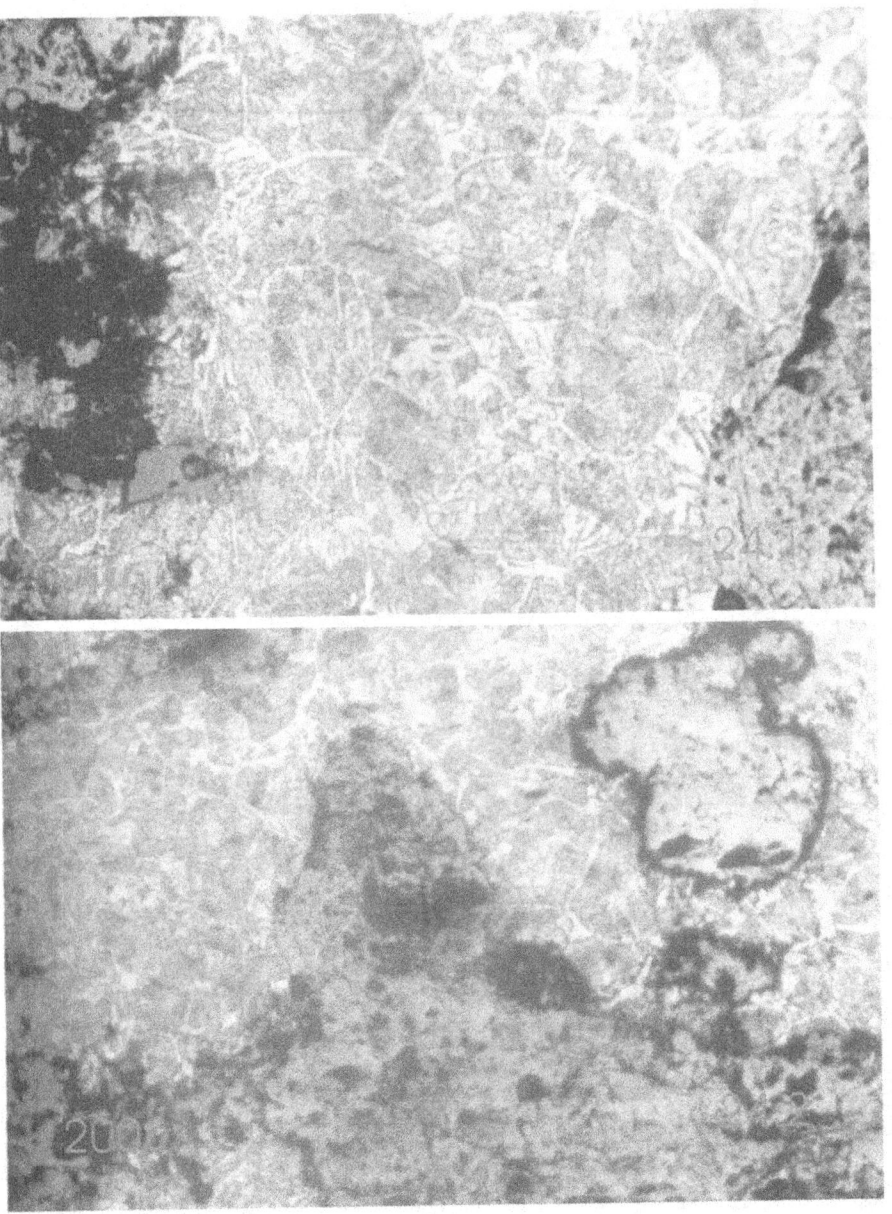

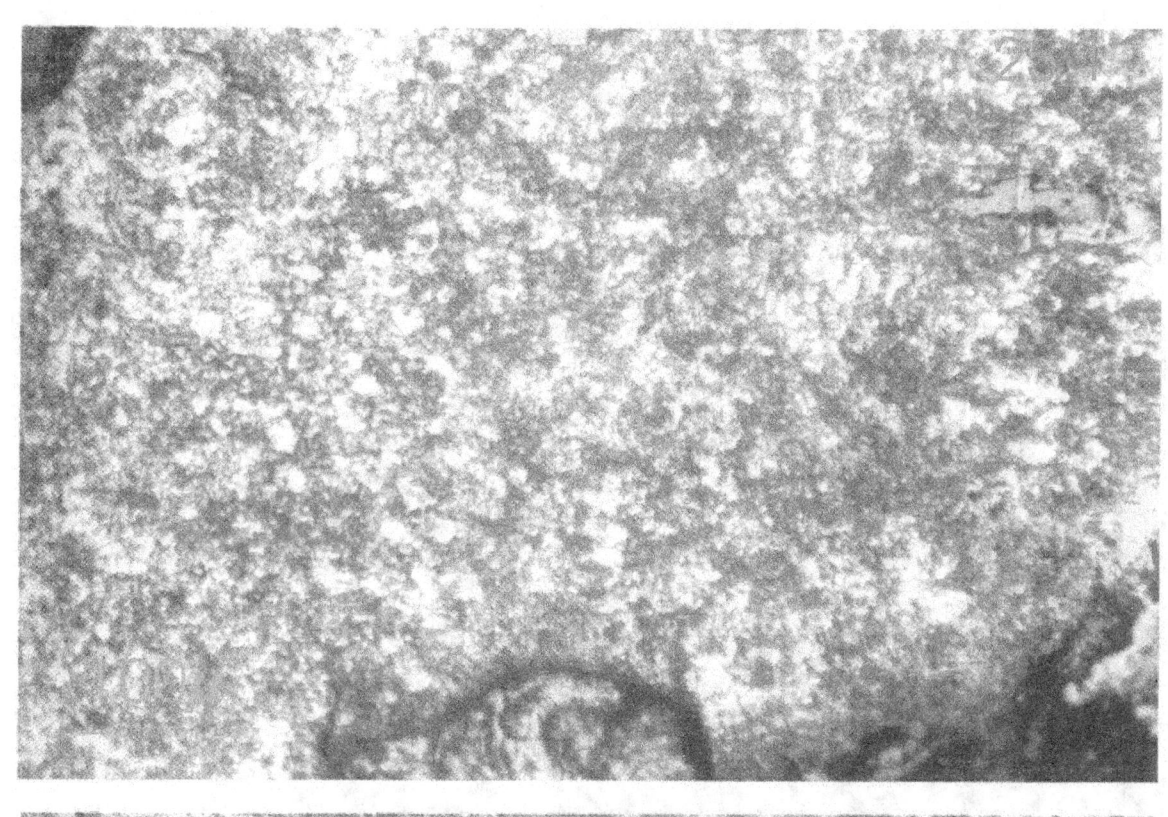

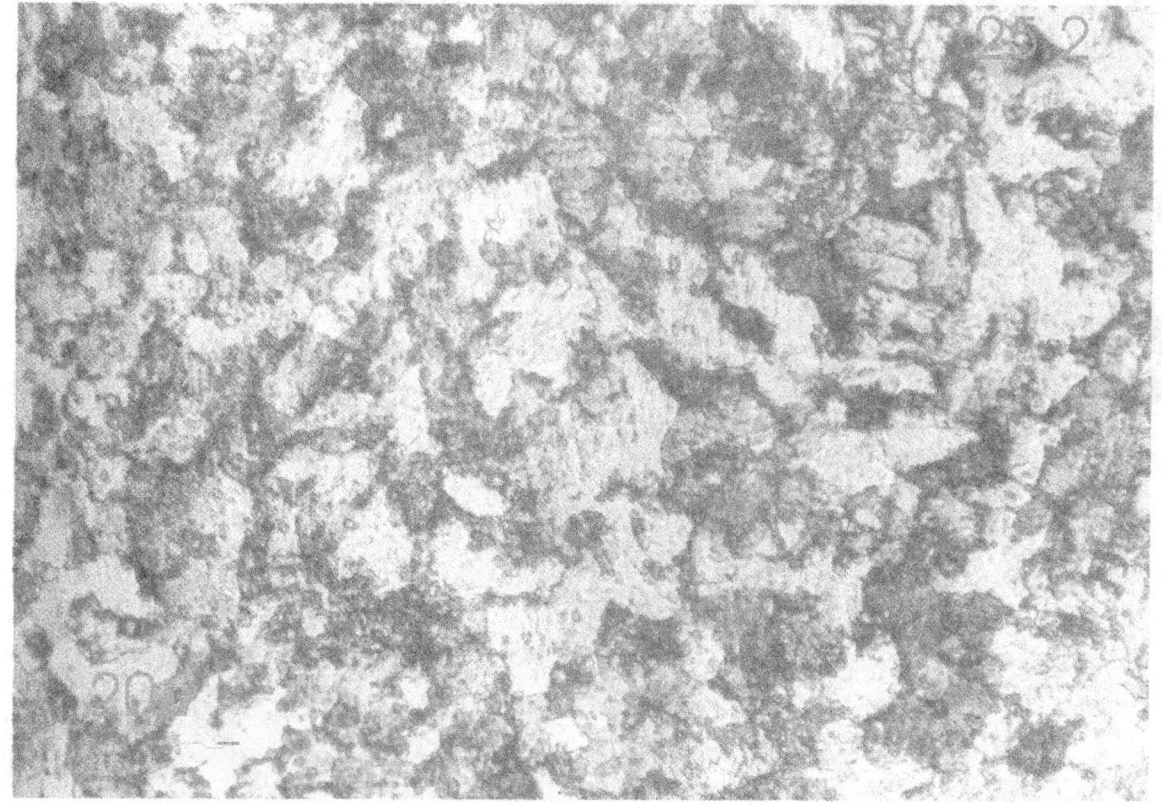

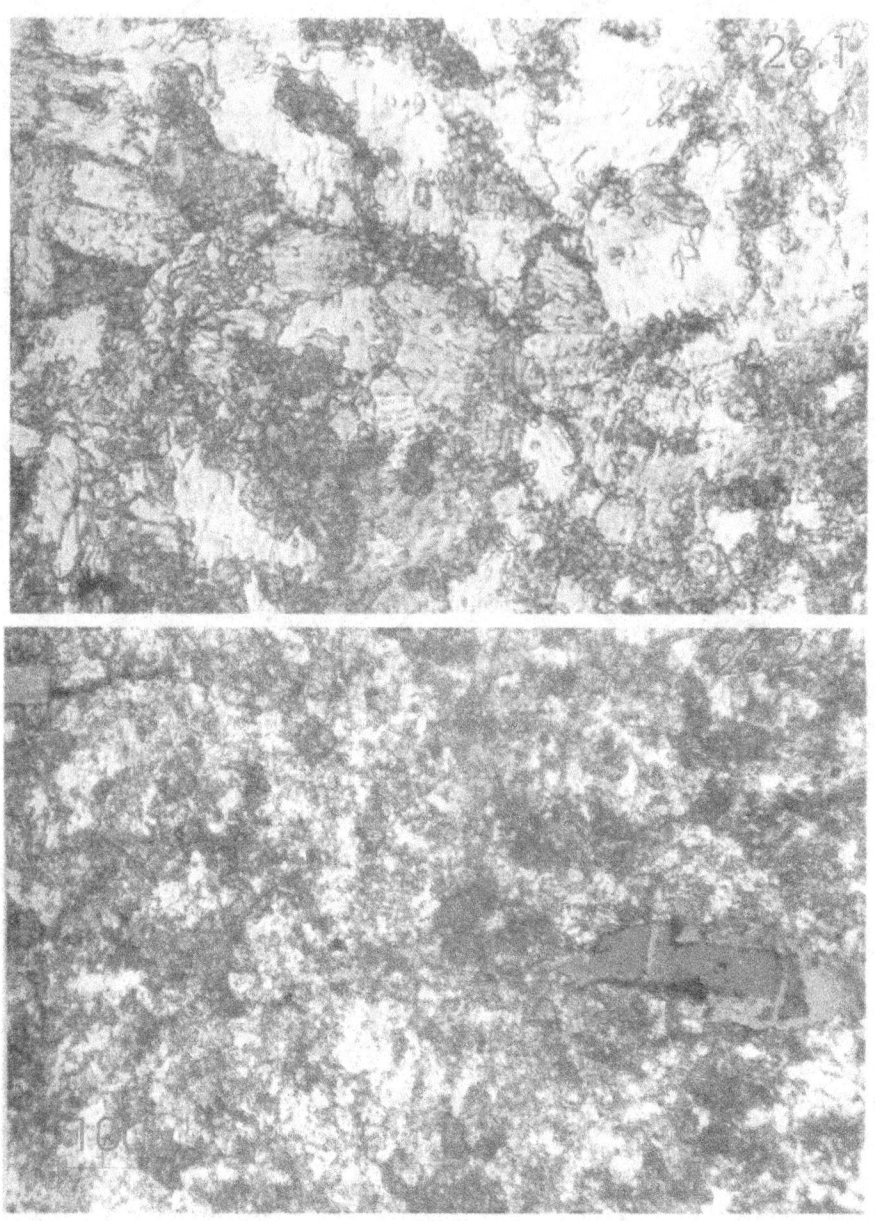

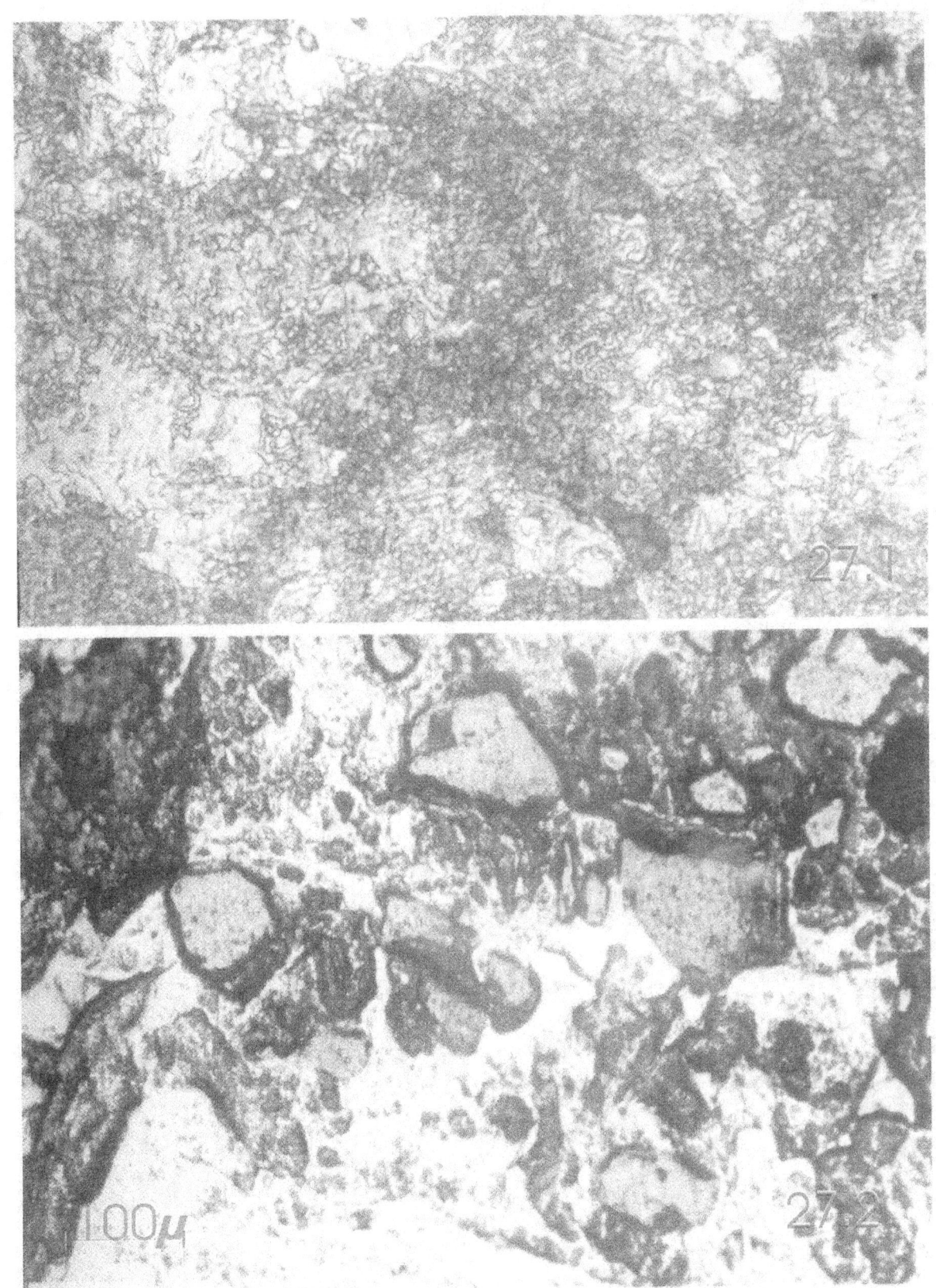

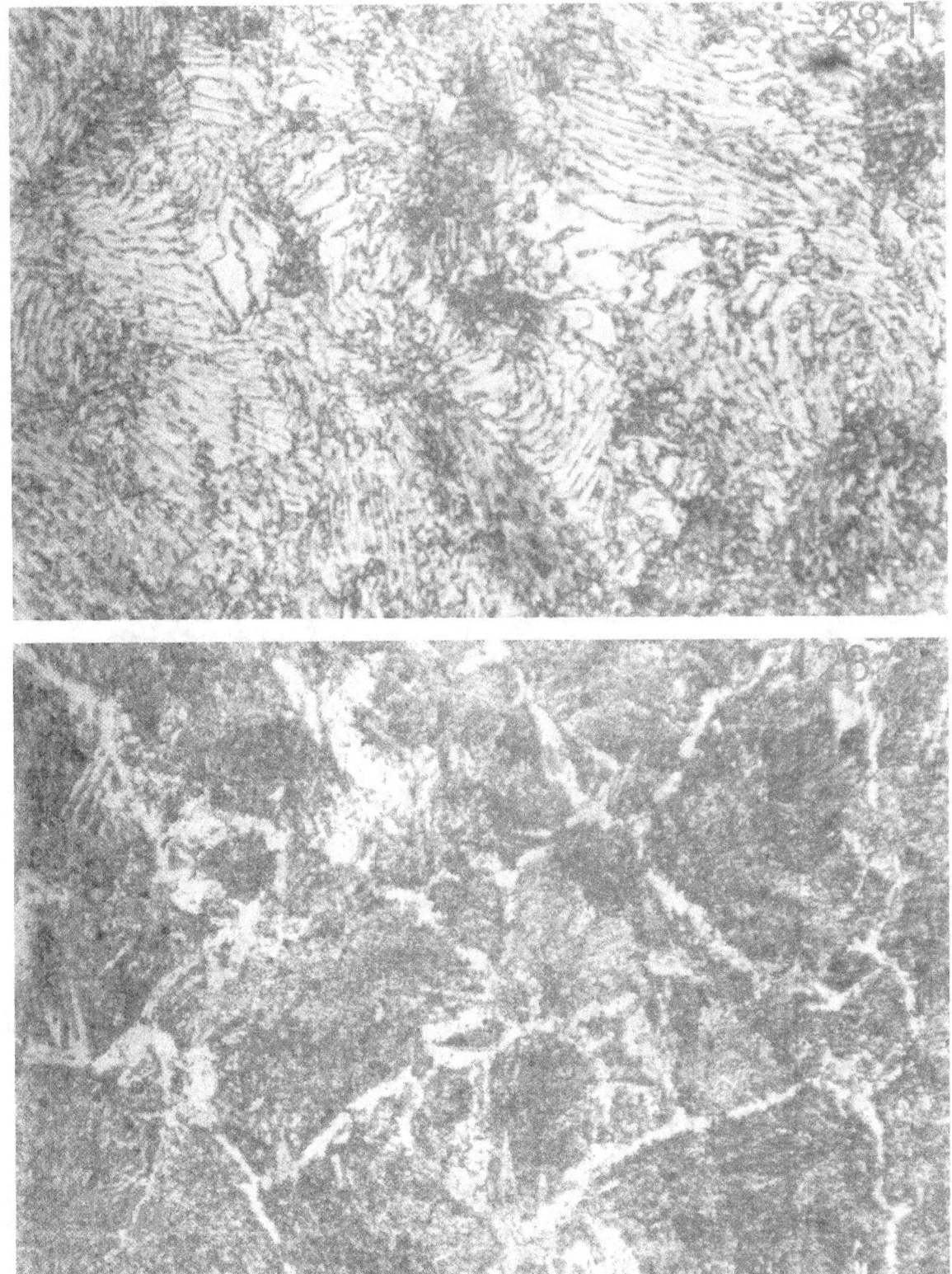

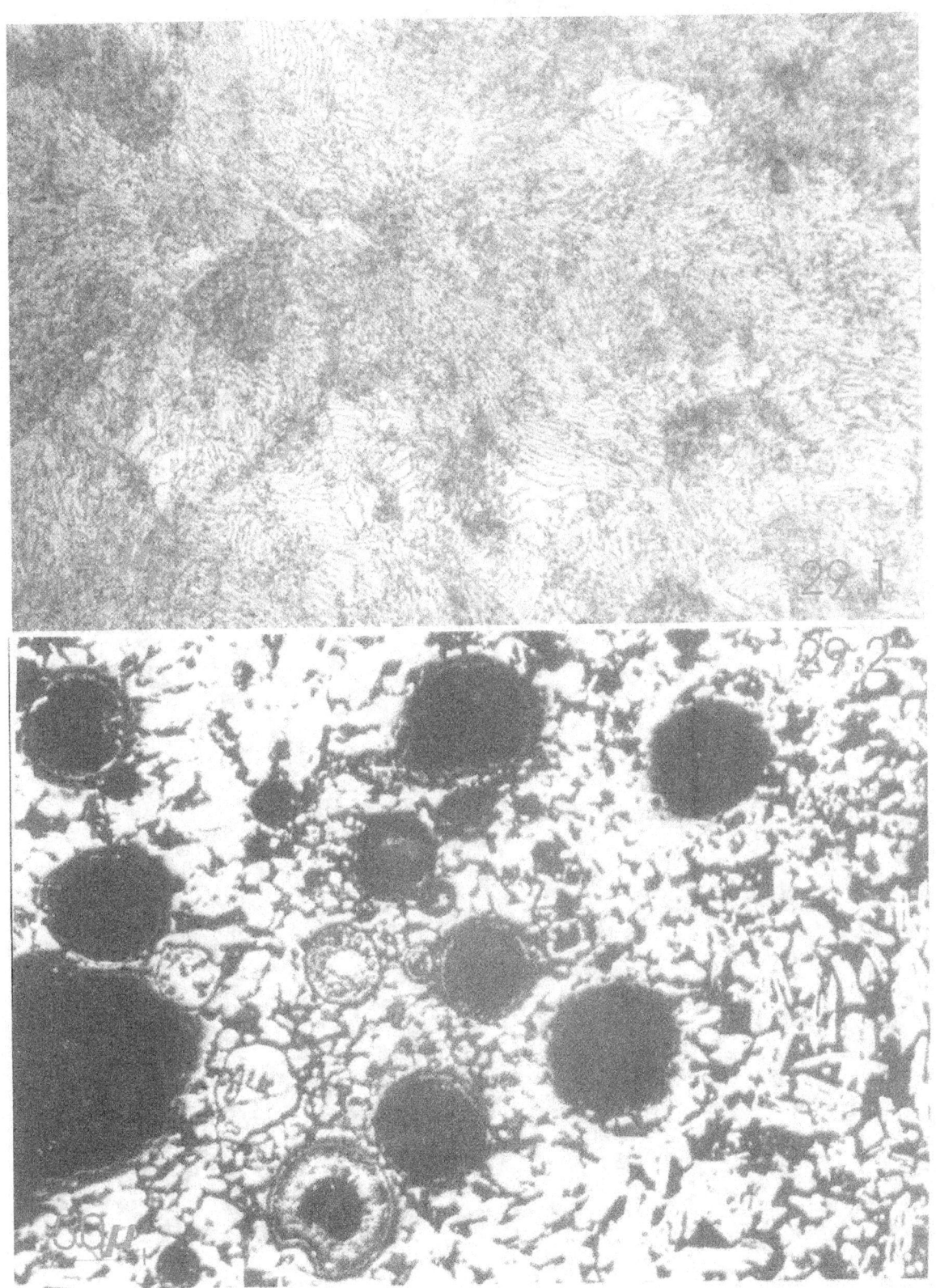

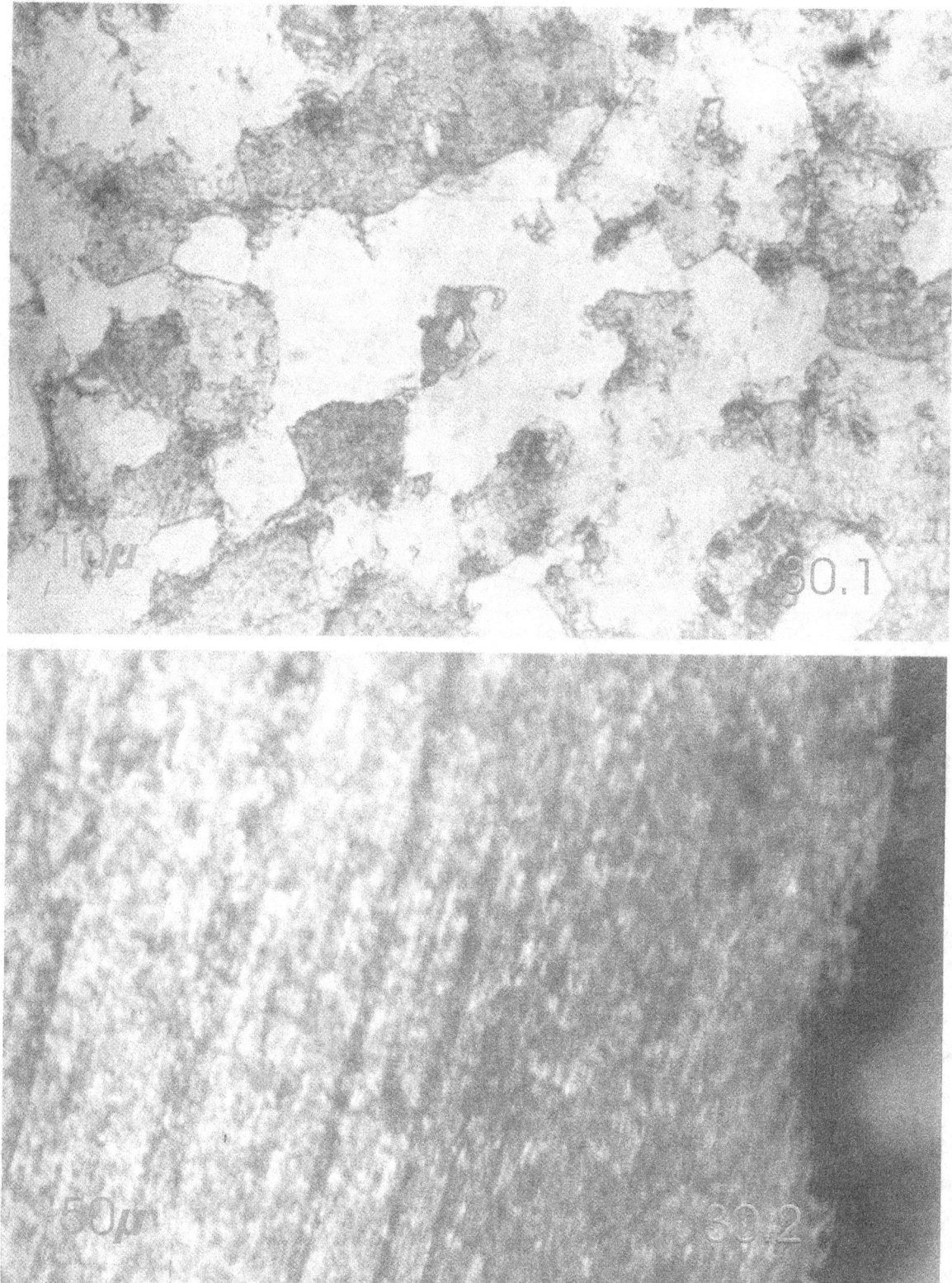

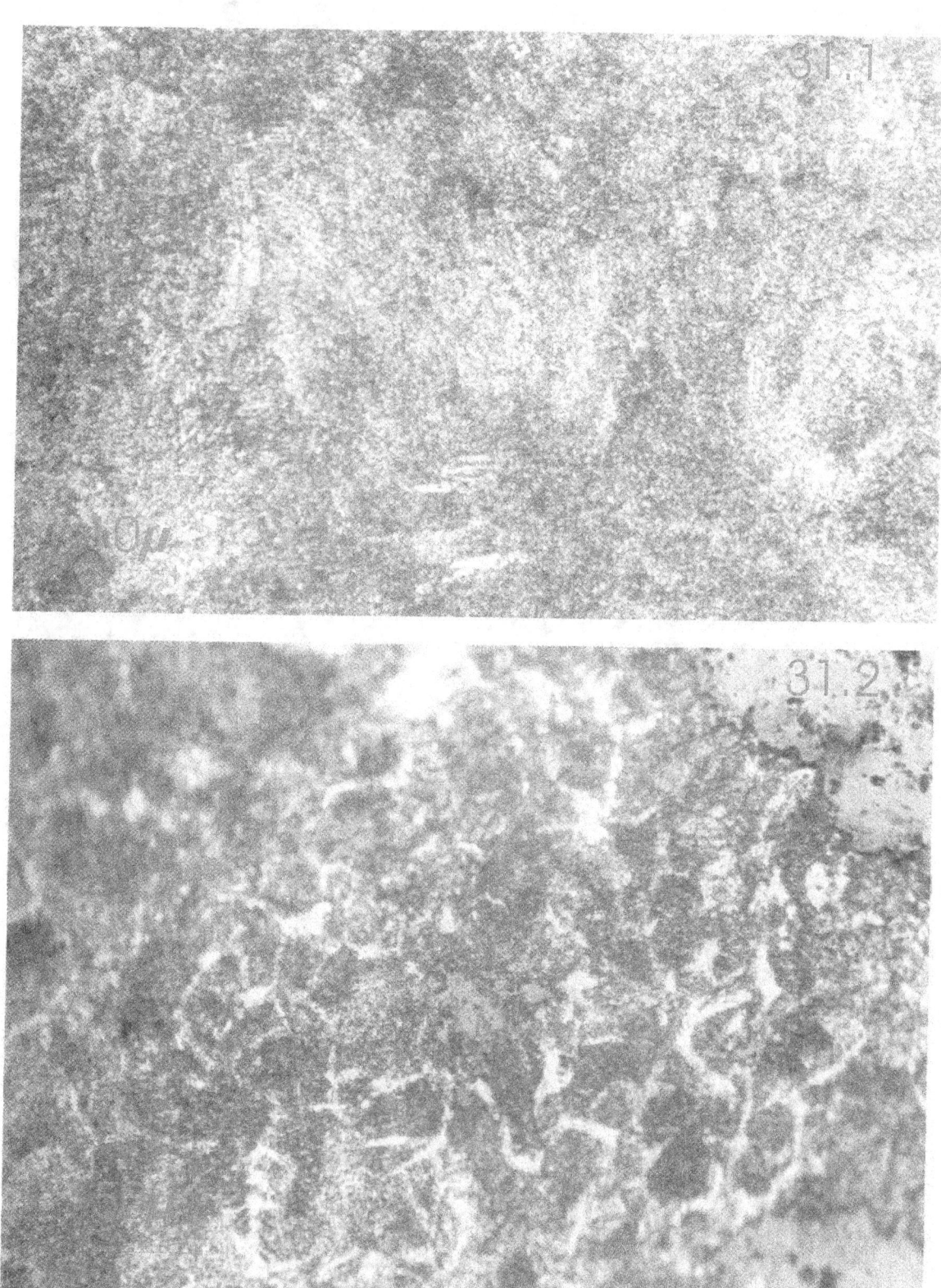

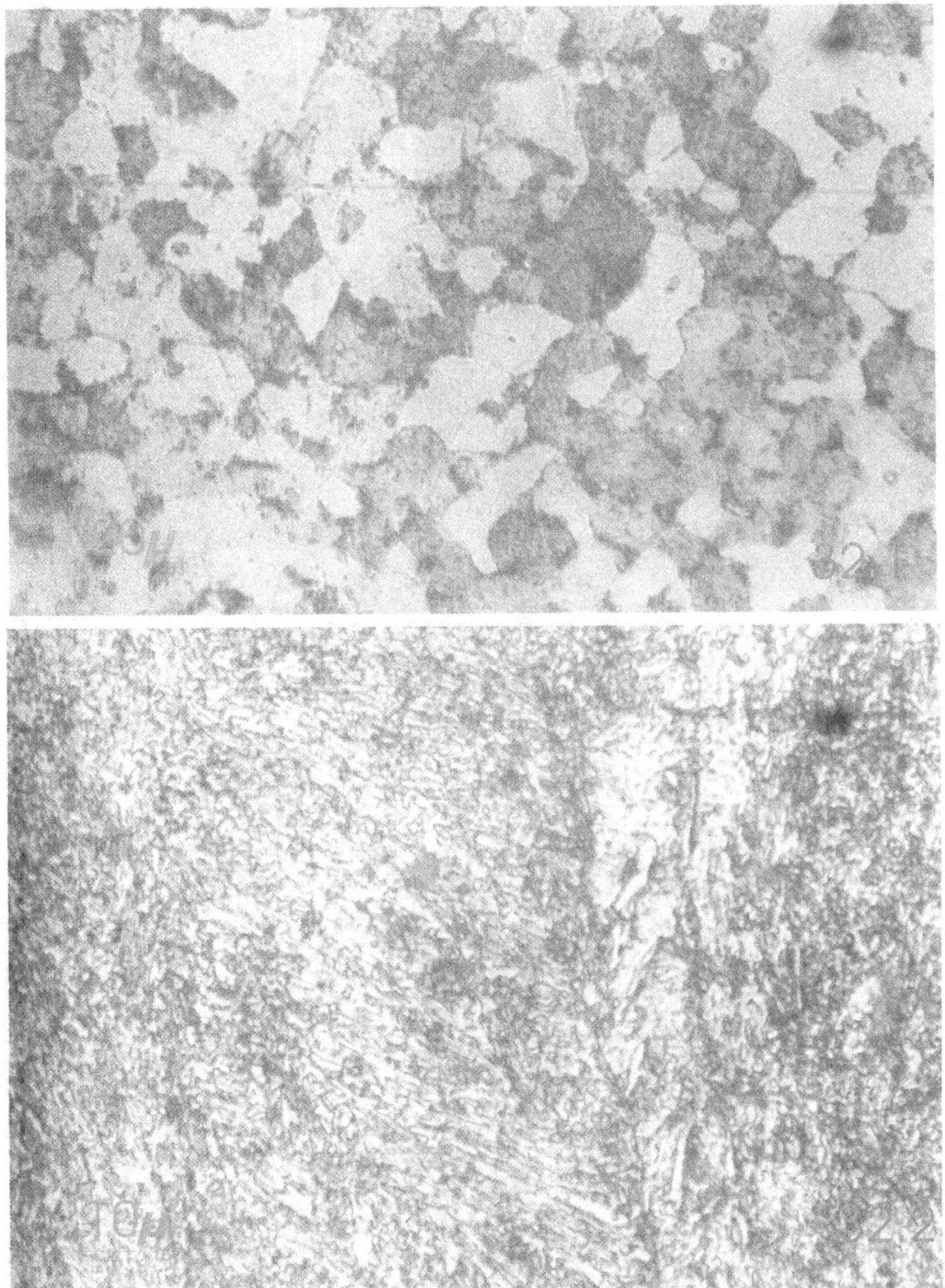

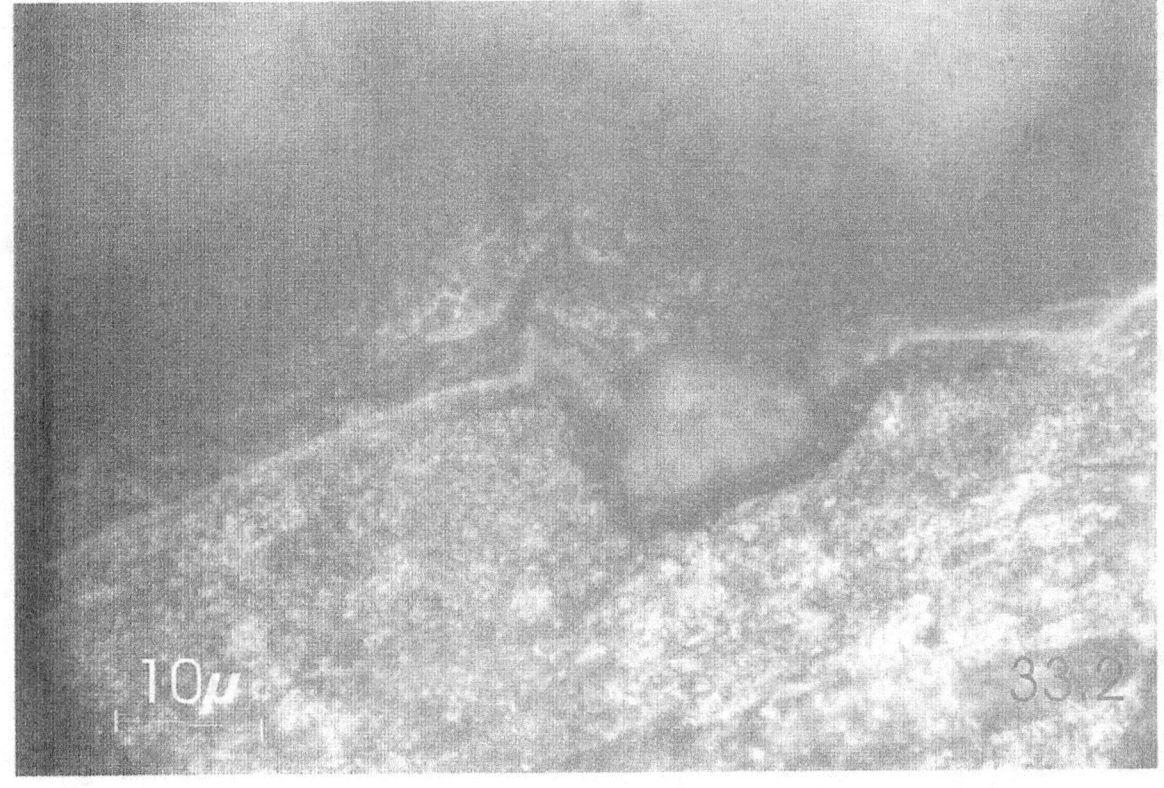

106 Iron Industry and Metallurgy: A Study

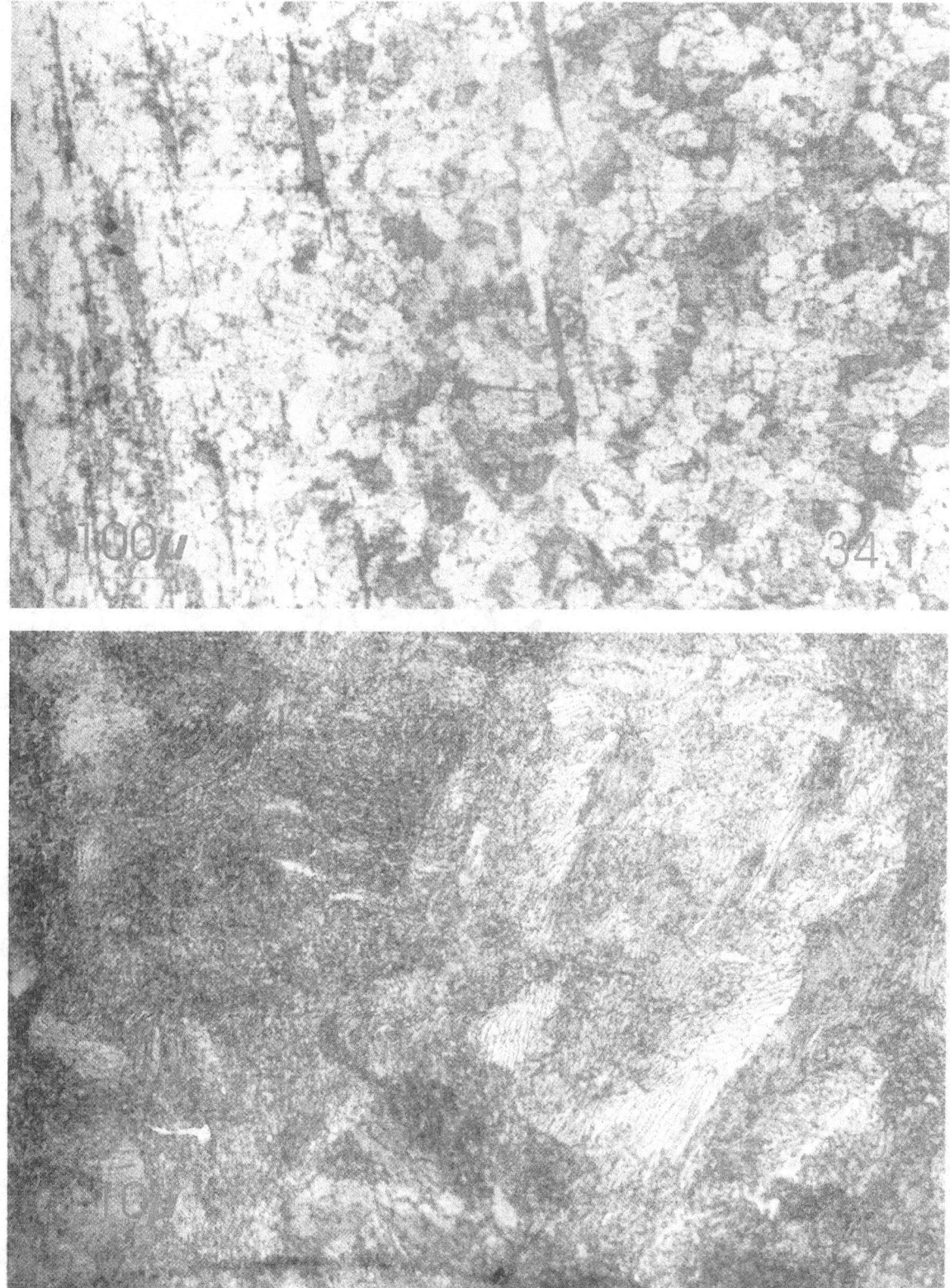

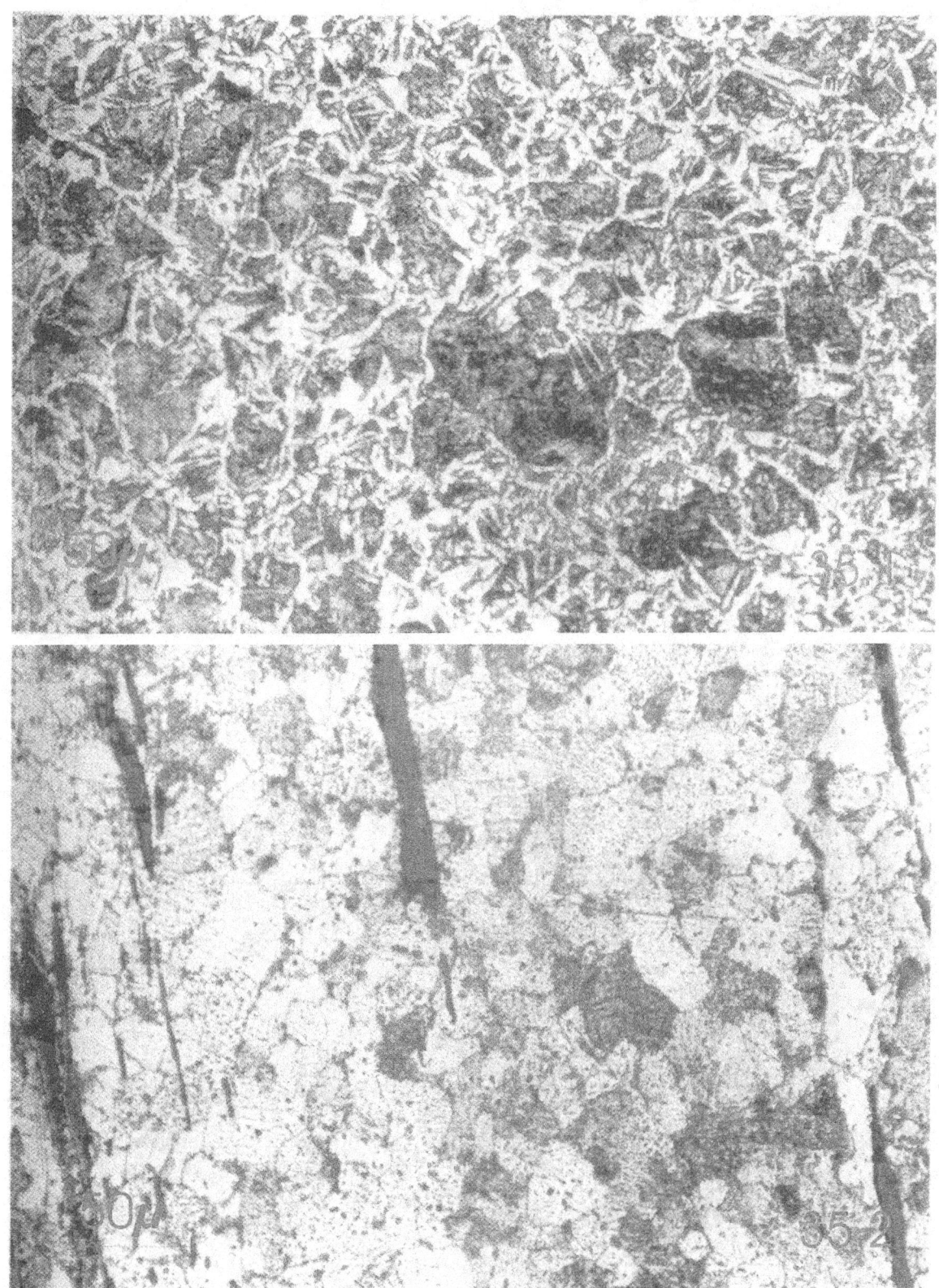

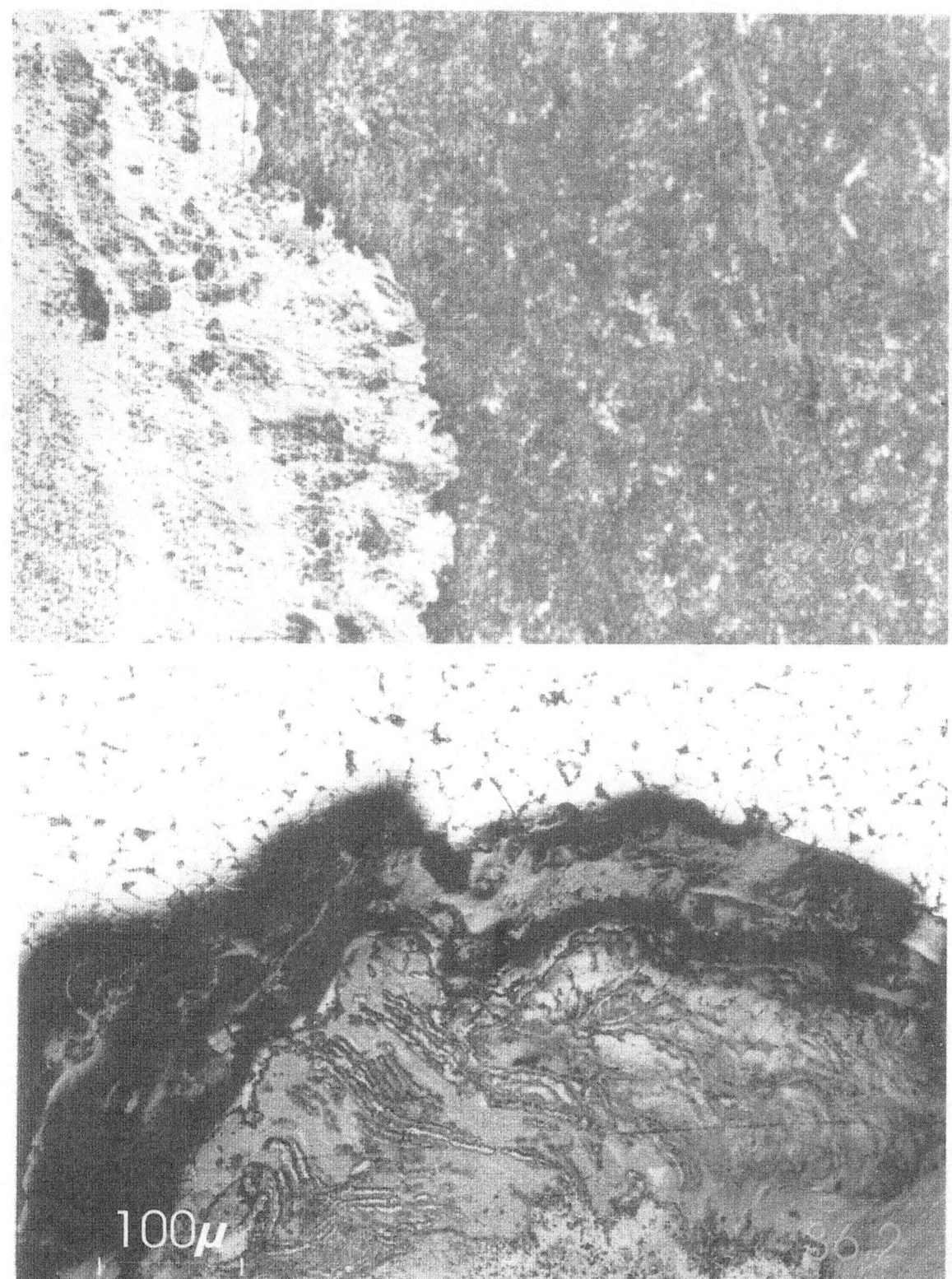

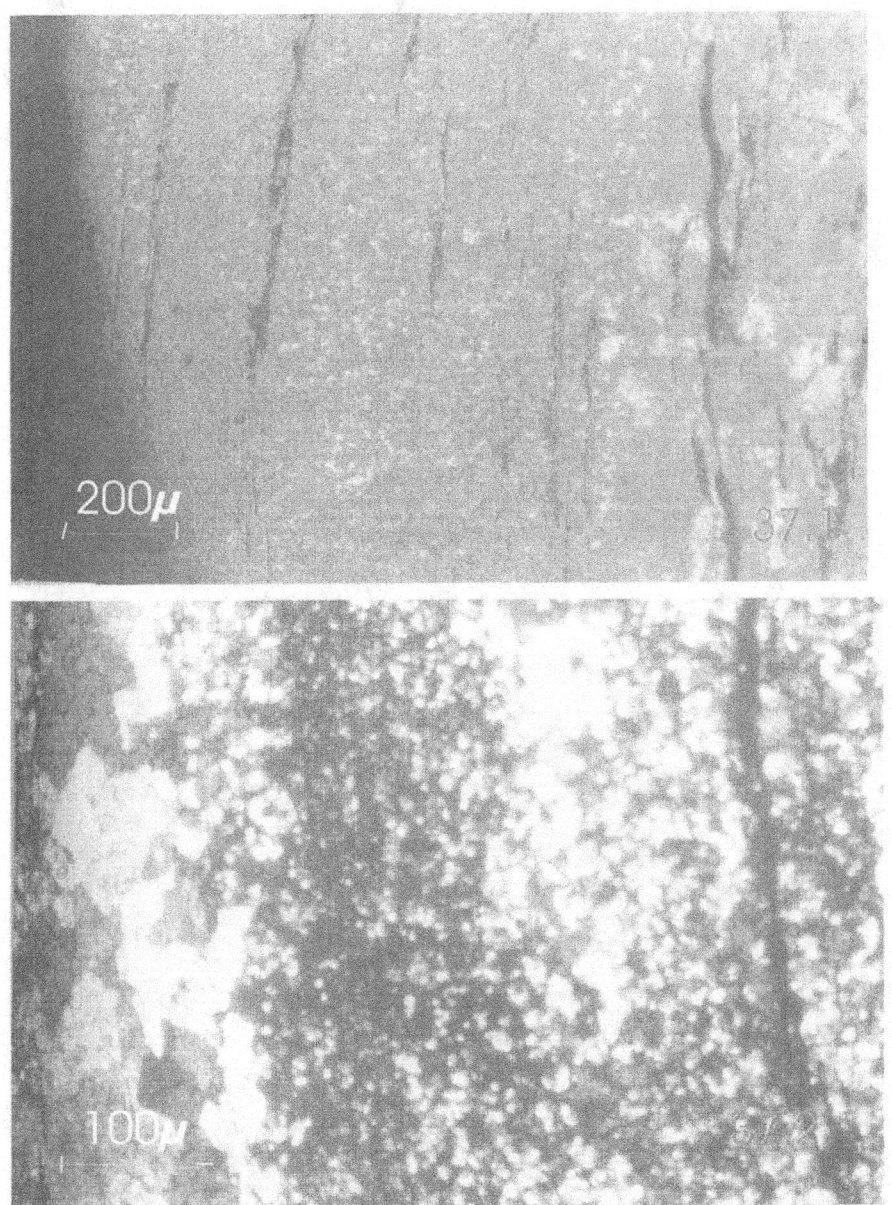

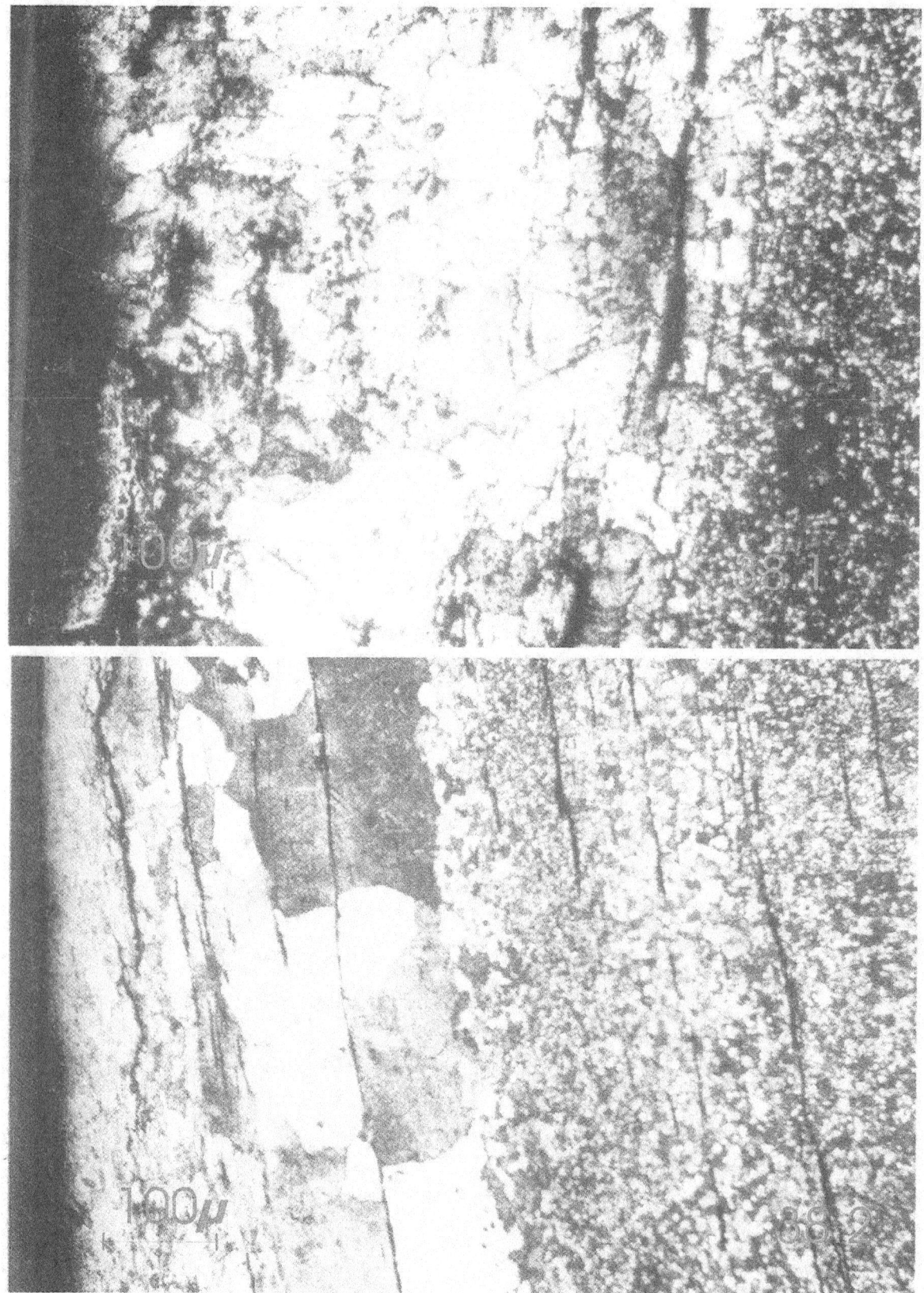

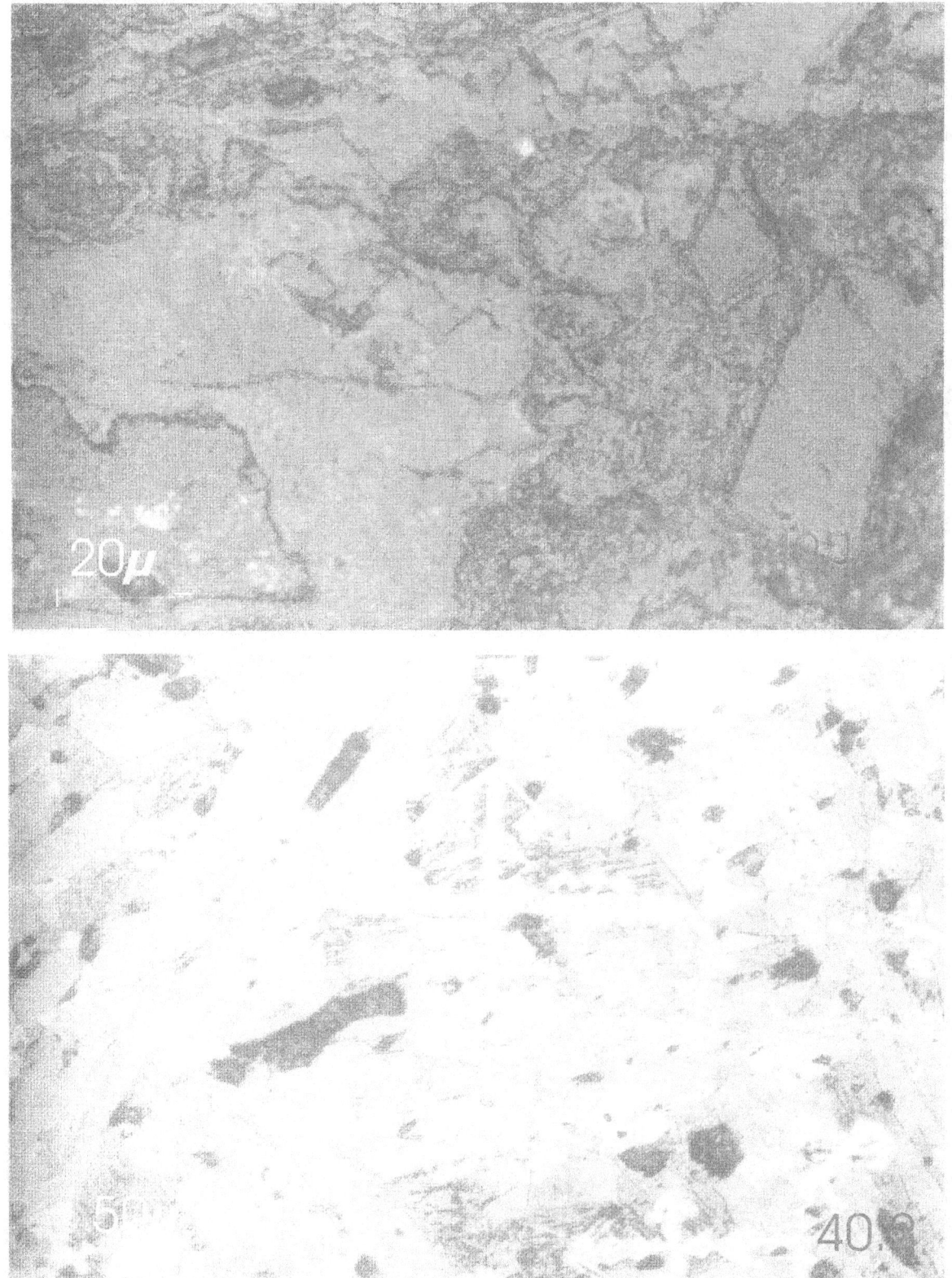

www.ingramcontent.com/pod-product-compliance
Lightning Source LLC
Chambersburg PA
CBHW080501220526
45465CB00006B/2335